SpringerBriefs in Economics

For further volumes:
http://www.springer.com/series/8876

Per-Olov Johansson · Bengt Kriström

Evaluating Water Projects

Cost-Benefit Analysis Versus Win-Win Approach

Springer

Per-Olov Johansson
Department of Economics
Stockholm School of Economics
Stockholm, Sweden

Bengt Kriström
CERE
SLU
Umeå, Sweden

ISSN 2191-5504 ISSN 2191-5512 (electronic)
ISBN 978-3-642-36789-2 ISBN 978-3-642-36790-8 (eBook)
DOI 10.1007/978-3-642-36790-8
Springer Heidelberg New York Dordrecht London

Library of Congress Control Number: 2013933019

Printed on acid-free paper

Springer is part of Springer Science+Business Media (www.springer.com)

Preface

The main purpose of this Brief is to present research on a kind of water use conflict that probably will become more and more common and important as time goes by. In times of increasing demands for electricity as well as environmental services, the question arises how to best manage moving water. Should more water be diverted to or from electricity generation? That is the kind of timely question which this Brief addresses.

Two different approaches are introduced and compared within two large empirical studies of river use. The first is a cost-benefit analysis of re-regulating a Swedish hydropower plant. Some water is currently diverted from electricity generation to the downstream dryway and we investigate a change of this regulation. The proposed scenario generates environmental and other benefits but comes at a cost in terms of electricity foregone. The analysis is complicated by issues such as foreign ownership of hydropower stations, renewable energy certificates, carbon emission permits, transmission of electricity provided by natural monopolies, externalities of replacement power generated in other countries, and so on. Our study can be seen as a kind of manual that can be used to evaluate reasonably small re-regulation of rivers.

The second study introduces an approach that is very different from the one used in a conventional cost-benefit analysis. Our approach provides a package of measures designed so that most, if not all, affected will be better off. Thus, in contrast to a conventional cost-benefit analysis which draws on hypothetical compensation measures, our new approach envisages actual compensation. While there is no monetary compensation, we propose to allow additional turbines at an existing plant, and "compensate" by other measures in the *same* river. We show that our proposed measures makes almost everybody better off, at least those living in the river catchment area. This Win–Win approach suggests a new way of handling difficult resource use conflicts that can be used in many other cases. Or so we will argue.

The research presented in this Brief was carried out with financial support from PlusMinus—Economic Assessment for the Environment—sponsored by the Swedish Environmental Protection Agency and Hydropower—Environmental Impacts, Mitigation Measures and Costs in Regulated Waters—financed by Elforsk, the Swedish Energy Agency, the National Board of Fisheries and the Swedish

Environmental Protection Agency. Many people from these organizations have contributed in discussions of different aspects of hydroelectricity generation. We are most grateful for these contributions. Several persons have helped us with the web surveys: Scott Cole, Kjell Leonardsson, Bo Ranneby, and Peter Rivinoja. Much of the necessary ecological work was undertaken by Kjell Leonardsson, for his contributions we are most grateful. We have benefited from detailed suggestions on the US experience within the field from John Duffield and V. Kerry Smith. Erik Brockwell checked the language. Any remaining errors are our own responsibility.

Finally, we are grateful to Springer-Verlag for allowing us to draw extensively in Chap. 2 of this Brief on our recent book. The Economics of Evaluating Water Projects—Hydroelectricity Versus Other Uses (ISBN 978-3-642-27669-9).

Stockholm and Umeå, February 2013

Per-Olov Johansson
Bengt Kriström

Contents

1 Introduction 1
1.1 The Ljusnan River and the Two Proposals at a Glance 3
1.1.1 The Dönje Hydropower Proposal 4
1.1.2 The −2,+1 Hydropower Proposal 7
1.2 Structure of the Brief 8
References 9

2 The Dönje Hydropower Scenario 11
2.1 The Basic Cost-Benefit Rule 11
2.2 The Impact of Changed Water Flow at the Hydropower Plant 16
2.3 Contingent Valuation Study of Improved Downstream River Basin 17
2.4 The Cost of Electricity Foregone 20
2.5 Pollution Externalities of Replacement Power 23
2.6 Social Discount Rate 23
2.7 Results 25
2.8 Sensitivity Analysis 27
2.8.1 Demand Changes 28
2.8.2 A Stochastic Sensitivity Analysis Based on Simulation Techniques 29
References 32

3 The −2,+1 Hydropower Scenario 35
3.1 A Sketch of the Theory Behind the Approach 35
3.2 Background 37
3.3 The Proposal and its Main Consequences 38
3.4 Web Survey 41
3.5 A Sum-Up of the Proposal's Impact 44
References 46

4 A Brief Comparison of the Approaches and an Outlook 49
References 54

Appendix A: A Simple General Equilibrium Cost-Benefit Rule 55

Appendix B: On a Win-Win Situation and the (Strong) Pareto Criterion 59

Appendix C: The Questionnaires 63

Index 65

Chapter 1
Introduction

Regulating a river alters flows and generally has strong negative impacts on most aquatic organisms and on those in the riparian zone (e.g. [1]), as well as detrimental effects on recreational possibilities and aesthetic values associated with rivers. On the other hand, hydropower offers (virtually) emission-free energy. In addition, hydropower is extremely cost-effective in countries with suitable natural conditions, like Norway and Sweden. Taken together, these facts suggests a conflict between competing uses of river resources in general and between policy objectives in particular. Consider the European Water Framework Directive, which formalizes the demand for improved ecological status of water bodies within the union in terms of quantified minimum levels. Simultaneously, the union has unleashed its "triple 20 by 2020" policy, which includes reducing carbon emissions by 20 percent. Furthermore, economic growth is a perennial policy objective and energy demand typically goes hand in hand with growth. These facts map into (increasingly sharper) conflicts about the proper husbandry of our water resources and, somehow, they must be resolved. This Brief offers tools that we hope can be useful in decision making involving resource use conflict.

Economists often rely on cost-benefit analysis (CBA) as one of those tools. It[1] offers a formal approach to delineating the costs and benefits of different policies and may provide useful information for decision-making. Quite arbitrarily, we refer the reader interested in reading more about the theoretical principles of project evaluation to [2–4]. These manuals are quite formal and demand some knowledge of general equilibrium theory. However, there are also many cookbook style manuals providing detailed advice on how to proceed in a real-world application, see e.g. [5–9]. An introduction to the underlying welfare theory is found in [10] while more technical presentations are provided by [4, 11].

Although the conceptual underpinnings of CBA may be traced to the nineteenth-century French civil engineer and economist Jules Dupuit, extensive application of the method had to wait until the twentieth century. The sharpening of CBA into a

[1] In Europe the approach is typically denoted cost-benefit analysis while in the U.S. it is often denoted benefit-cost analysis. We will throughout follow the European tradition in this respect.

P.-O. Johansson and B. Kriström, *Evaluating Water Projects*,
SpringerBriefs in Economics, DOI: 10.1007/978-3-642-36790-8_1,

potentially useful decision-making tool also involved engineers, i.e., the U.S. Army Corps of Engineers (ACE). Propelled by a rising demand for electricity and substantial damage from several serious floods, the U.S. Congress passed two significant flood control acts (1936, 1944) (referred to here as the 1936 Act and the 1944 Act, respectively).[2] The 1936 Act called for "works of improvement" on more than 50 major rivers throughout the United States and made flood control a federal government activity. The necessary physical constructions had been the ACE's expertise, and ACE became heavily involved in many construction projects, as it was given responsibility for analysis of rivers for flood control (whereas the U.S. Department of Agriculture was given responsibility for water flow on upstream watersheds).[3]

Importantly, the Act introduced an approach to prioritizing between projects:

> The federal government should improve or participate in the improvement of navigable waters or their tributaries, including watersheds ... for flood-control if the benefits to whomsoever they may accrue are in excess of estimated costs

(as quoted in [14]). These ideas were later developed in several handbooks and manuals. The 1944 Act gave the Corps responsibility for multi-purpose dams, e.g. hydropower constructions.[4] CBA subsequently conquered new worlds and new applications in the 1950s and onwards, as the tool was applied to various types of public projects in Europe and later on in the third world countries.

One of the studies we present in this Brief draws on these theoretical developments and provides a state-of-the-art cost-benefit rule of re-regulating the Dönje hydropower plant[5] on the river Ljusnan in mid Sweden. Water is redirected from electricity generation to the dryway. This will generate recreational, ecological, and other benefits but there is of course a cost in terms of electricity foregone. The proposed re-regulation will be described in some detail below.

The second study considers another re-regulation but from a very different perspective; it forms the basis for our second approach. The basic idea is a (rather complex) re-regulation such that, at best, everyone becomes a winner. Two dams at the mouth of the river Ljusnan are removed. This will allow salmon to migrate some 150 km to the pre-regulation natural barrier. As a compensation the hydropower operator is allowed to install additional turbines underground at the existing Laforsen plant (constructed at the waterfall that served as the natural barrier before river regulation). By scenario construction, the downstream winter water flow will basically mimic the before-regulation flow. In simple terms, we propose to re-direct the "excess winter flow" in a long tunnel such that we can restore the pre-regulation conditions

---

[2] According to [12], the federal Reclamation Act of 1902 required economic analysis of projects.

[3] For further details on the problems created by this separation of tasks, see [13].

[4] For a historical review of the development of CBA in the U.S. the reader is referred to [15], for Australia to [16], and for some UK studies to [17]. A fine book-length treatment of water planning in the U.S. that covers the developments in great detail is the volume by [18], see, especially, Table 3.1 in the chapter by [19] for a comprehensive overview of the developments.

[5] All hydropower stations considered in this Brief are owned by Fortum, a multinational quoted on NASDAQ OMX Helsinki, Finland, with a turnover of EUR 6 bn in 2011 and ranked 13th largest European electricity supplier.

in the main stream. From a bystander's point of view, the river will appear to have reverted to its pre-regulatory flow (the increase in electricity generation is, of course, not visible). In a way, the scenario reconstructs the natural winter flow, while simultaneously increasing electricity generation; this is a win-win regulatory change. Or so we will argue.

In a sense, this *Win-Win* approach is consistent with the one suggested by Nobel Laureate James Buchanan.[6] Somewhat simplified, he suggested that a measure should be undertaken only if everybody is better off with it; see, for example, [20]. However, in our full proposal, which includes additional measures explained in the next section, there is no monetary compensation; a package of measures is designed such that, at best, everybody is better off. This is in sharp contrast to the typical cost-benefit analysis which draws on the Hicks-Kaldor criterion—a proposal is recommended if those who gain from it are able, at least *hypothetically*, to compensate those who lose from it.

Let us now turn to a more detailed presentation of our proposals. While they are specific to the river Ljusnan, the basic ideas have wide applicability as we will explain in Chap. 4.

1.1 The Ljusnan River and the Two Proposals at a Glance

The Ljusnan River in mid/northern Sweden, see Fig. 1.1, originates from the mountains at the Norwegian border and has a catchment area of about 19 800 km^2 which along with its hydropower plants are indicated in Fig. 1.2. It is around 430 km long and has an annual mean flow of $226\,\text{m}^3\text{s}^{-1}$ at the mouth in the Bothnian Sea.[7]

Ljusnan has long been used as a power source, but the main construction phase began in the 1950s. Nowadays the river is regulated along most of its length for hydropower production except that a 50 km long section, Middle Ljusnan, has been protected from development. In total there are more than 25 power plants spread out over the river and its small tributary Voxnan (there are also a number of mostly very small (<1 megawatt) plants spread out over the catchment area). These are indicated in Fig. 1.2 but only the plants under investigation in the present study are named. The typical annual production amounts to some 3.8 TWh.

Salmon fishing in Ljusnan is documented since the time of King Gustav Vasa in the 16th century. However, as a consequence of the construction of hydropower plants, salmon and other fish species are prevented from passing beyond the first dam on the coast at Ljusnefors; upstream migration is limited due to extensive stretches of dryways.

[6] October 3, 1919–January 9, 2013.

[7] The Bothnian Sea and the Bothnian Bay constitute the Gulf of Bothnia, which is depicted in Fig. 1.1.

Fig. 1.1 Sweden with Ljusnan River in the *red rectangle. Source* http://www.worldatlas.com/webimage/countrys/europe/se.htm

1.1.1 The Dönje Hydropower Proposal

Dönje power station is a 72 megawatt[8] (MW) hydroelectric facility on the Ljusnan river close to the small city of Bollnäs in central Sweden. The city of Bollnäs has around 13 000 inhabitants and the municipality of Bollnäs has around 26 000 inhabitants.

The plant is depicted in Fig. 1.3 and its location on the Ljusnan River is seen from Fig. 1.2.

The river section Bollnäsströmmarna, c. 6.5 km in length, is located between power-station five and six (counting from the river mouth), about 53 km upstream of the sea. Generally speaking, most of the water reaching this section is utilized by the power-station Dönje. It has a maximum capacity of $250\,m^3s^{-1}$ with the hydraulic head created by 34 m deep intake tubes to the turbines, after which the water is lead via a tunnel to the outlet in the downstream Lake Varpen.

---

[8] According to owner Fortum. Its normal annual production is 340 gigawatt hours (GWh) according to [21].

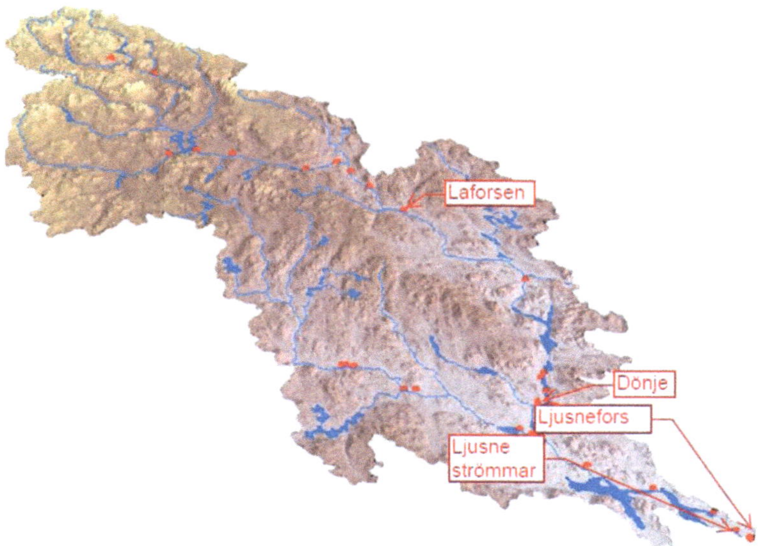

Fig. 1.2 The catchment area of River Ljusnan. Its hydropower plants are marked in *red* and the four under investigation are named. *Source* own work

Since water is diverted from the natural river stretch this results in reduced flows in Bollnäsströmmarna. According to the hydropower licence conditions, the minimum flow in the original stretch is limited to $10\,m^3s^{-1}$ from May 15 to October 21, after which the spill is gradually decreased to $0.5\,m^3s^{-1}$ for the winter period (October 31 to May 14). In the summer, most of the water is spilled into the 1.3 km long eastern river branch, Klumpströmmen, that receives $0.25\,m^3s^{-1}$ in winter, while the western branch is more or less dry for most of the year. Our scenario targets Klumpströmmen, see Fig. 1.4 for a map.

Before regulation, about one third of the total flow in Bollnäsströmmarna went into the Klumpströmmen section, while the remaining 2/3 passed the western river branch. The natural flow regimes showed the typical patterns for rivers in northern Sweden; the highest flow events occurred in spring (peaking c. $740\,m^3s^{-1}$), with decreasing and stabilizing flows in the summer (normally c. 15–70 m^3s^{-1}), occasionally with an increase in autumn, and relatively low constant flows during the ice-period in winter (averages around $10\,m^3s^{-1}$). Today, the dryway is more or less bottom frozen during the winter. The downstream stocks of fish valued for fishing are small. The scenic view when the river channel is dry is far from overwhelming. Recreational activities other than fishing, such as canoeing and ice skating, are also adversely affected by the regulation.

Two scenarios were described in our websurvey. These were developed via focus group studies and in-depth discussions between various stakeholders, ecologists and economists. In the first scenario, the winter flow would increase from $0.25\,m^3s^{-1}$ in Klumpströmmen to $3\,m^3s^{-1}$ while the summer flow remained unchanged at its

Fig. 1.3 The Dönje Hydropower plant. *Source* Unknown

current level of $10\,\mathrm{m}^3\mathrm{s}^{-1}$ per second. In the second scenario we also increased summer flow, from 10 to $20\,\mathrm{m}^3\mathrm{s}^{-1}$. In this Brief we will stick to the first scenario.

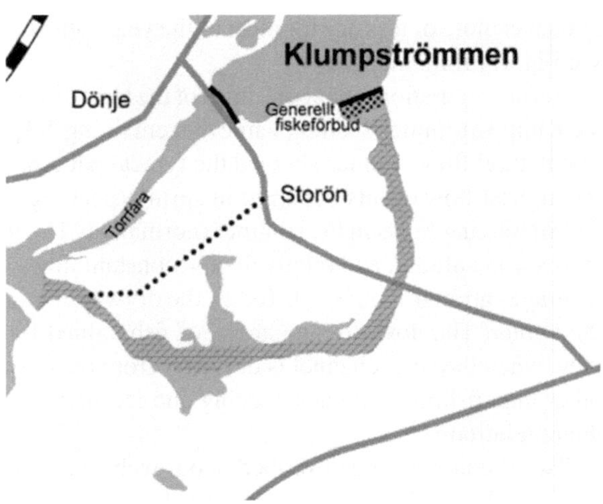

Fig. 1.4 Map of Klumpströmmen at the power plant. Note: "Torrfåra" = dry river bed, "Generellt Fiskeförbud" = Fishing not allowed. *Source* Municipality of Bollnäs

This is motivated by the fact that there is a difference in magnitude, but not "quality" between the scenarios (our respondents echoed this sentiment).[9]

Given the scenarios, we then proceeded to estimate benefits and costs. We used a small contingent valuation web-questionnaire to measure the benefits and a meta-study to estimate the environmental costs of replacement electricity; in a worst-case scenario, the hydropower foregone will be replaced by coal-fired plants abroad. The expected loss of present value profits due to reduced electricity generation is also estimated requiring, inter alia, a forecast of future electricity prices. We will come back to the details of estimating these items in Chap. 2.

1.1.2 The −2,+1 Hydropower Proposal

The other project that is considered in this Brief is more complex than the Dönje project since it consists of two parts and involves 3 dams. According to the proposal two dams at the end of the river, Ljusne strömmar (Ljusne Currents) and Ljusnefors, see Fig. 1.2, are removed. This part of the proposal is denoted "−2" since 2 dams are removed and results in an estimated annual loss of some 393 GWh. The other part of the proposal, denoted "+1", implies that the plant at Laforsen is refitted with new underground turbines and some $50\,m^3s^{-1}$ are diverted to the natural riverbed. The remaining around $100\,m^3s^{-1}$ are used to generate some 470 GWh per annum. The +1-part of the proposal is outlined in Fig. 1.5. The project is thus expected to result in a net increase in electricity generation. In addition, the project would allow salmon to migrate up to the natural barrier at Laforsen, while Ljusnefors at the mouth of the river is the barrier today. It would also provide other environmental and recreational benefits. For example, the winter flow of water in Middle Ljusnan will roughly mimic the one before the river was regulated. Therefore, it provides a kind of Win-Win option where almost all affected parties, including the operator of the hydropower plants and those living in the river basin, are expected to win.

A web survey is used to shed light on the attitudes of those in the catchment area of the river affected by the proposal. The idea is to test the hypothesis that the inhabitants are no worse off with the proposal than without it. Own calculations of the present value revenues and costs of the change in electricity production are used to check if the owner of the hydropower plants (the multinational Fortum) is no worse off with the proposal. If we get affirmative answers on both tests we could argue that the proposal, if implemented, provides a Pareto improvement (provided at least someone is strictly better off); recall that the (strong version) of the Pareto criterion is satisfied if someone is better off and no one is worse off with a proposal/project.

---

[9] From an ecological point of view the optimal flow mimics the natural flow. A "natural flow" scenario was considered but not implemented in the study due to its complexity.

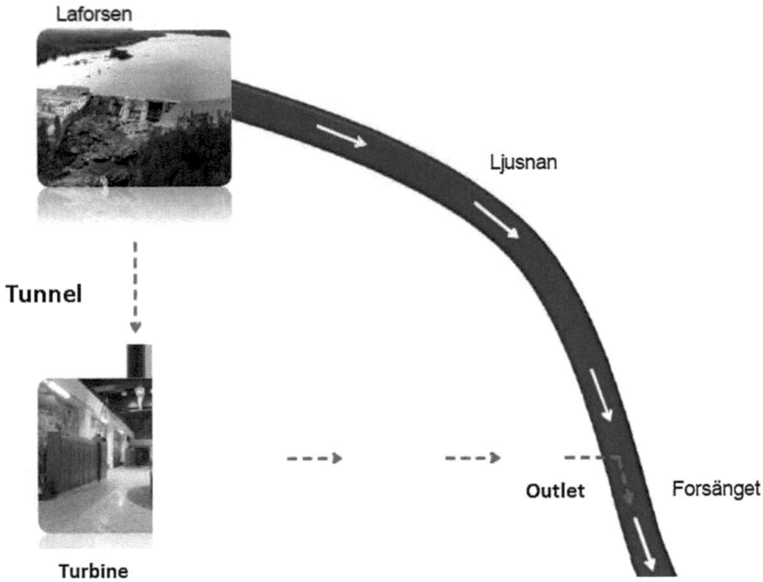

Fig. 1.5 The "+1" part of the "−2,+1" proposal. *Source* own work

1.2 Structure of the Brief

Chapter 2 of the Brief sums up our general equilibrium cost benefit analysis in [23] of a proposed re-regulation of the Dönje hydropower plant in the Swedish river Ljusnan. An introductory section is devoted to a sketch of the general equilibrium cost-benefit rule used to assess the project. Although the rule is astonishingly simple, the underlying model of a small open economy is quite complex. Sections 2.2 and 2.3 of the chapter provide a brief overview of the contingent valuation study that is used to assess the benefits of the considered re-regulation. The cost of the proposal is the loss of electricity and hence loss of profits for the hydropower plant owner. Section 2.4 sums of the different items in the profits loss expression. In addition, replacement electricity produced by coal-fired plants cause an externality cost, e.g. emissions of climate and other gases. The associated cost estimate is presented in Sect. 2.5. The choice of discount rate is discussed in Sect. 2.6. The point estimates of the different benefit and cost terms are contained in Sect. 2.7. Finally, a brief sensitivity analysis is undertaken in order to check the robustness of the point estimates.

Chapter 3 is devoted to what we believe is a brand-new approach to empirical assessments of re-regulations. We name the study −2,+1 because it involves the removal of two dams (−2) close to the mouth of Ljusnan River and the addition of new turbines upstream (+1). The −2-part of the proposal would allow salmon to migrate, as noted, some 150 km upstream to the natural barrier. There are also recreational and aesthetical benefits associated with the proposal. The owner of the removed dams is allowed to install new turbines at the Laforsen plant(+1). Overall, the owner is expected to gain from the proposal. Therefore, the hypothesis is that (almost) everyone gains from the proposal. In this sense, it is similar to Nobel laureate James Buchanan's proposal from the late 1950s that only such projects should be undertaken. There is a sharp difference between this approach and a conventional cost-benefit analysis. In the latter some typically gain while others lose from a project (and compensation is not actually paid). We base the evaluation on a web-survey to people living in the river basin and estimate the benefits and costs of removing 2 dams and establishing a new plant. Thus we test the hypothesis that the proposal results in a "Win-Win" situation.

In Chap. 4 we contrast and compare these two approaches to handling natural resource conflicts. We do believe that our win-win approach used in Chap. 2 can be applied to many types of difficult natural resource use conflicts, where different and opposing interest groups are involved. Therefore, the approach should be of general interest. Such conflicts are, as noted, becoming increasingly common and sharper over time.

Appendix A provides a simple illustration of a general equilibrium cost-benefit rule of the kind used in the evaluation of the re-regulation of the power plant at Dönje, Appendix B provides an illustration of the theoretical foundations of the win-win approach used in the evaluation of the −2,+1 scenario, and Appendix C sketches the web-questionnaire that constitutes the basis of the evaluation in Chap. 3.

References

1. Saltveit, S.J.: The effects of stocking Atlantic salmon, Salmo Salar, in a Norwegian regulated river. Fish. Manag. Ecol. **13**, 197–2005 (2006)
2. Drèze, J., Stern, N.: The theory of cost-benefit analysis. In: Auerbach, A., Feldstein, M. (eds.) Handbook of Public Economics. North-Holland, Amsterdam (1987)
3. Johansson, P.O.: Cost-Benefit Analysis of Environmental Change. Cambridge University Press, Cambridge (1993)
4. Myles, G.D.: Public Economics. Cambridge University Press, Cambridge (1995)
5. de Rus, G.: Introduction to cost-benefit analysis. Looking for reasonable shortcuts. Edward Elgar, Cheltenham (2010)
6. European Commission: Guide to cost-benefit analysis of investment projects. Technical report, DG Regional Policy (2008)
7. HM Treasury: The green book. Appraisal and evaluation in central government. HMSO, London. http://greenbook.treasury.gov.uk/index.htm (2003)
8. Pearce, D., Atkinson, G., Mourato, S.: Cost-benefit analysis and the environment. Recent developments. Organization for Economic Co-operation and Development, Paris (2006)
9. EPA: Guidelines for Preparing Economic Analyses. U.S. Environmental Protection Agency (EPA), Washington, DC. EPA 240-R-00-003 (2000)
10. Johansson, P.O.: Introduction to modern welfare economics. Cambridge University Press, Cambridge (1991)
11. Boadway, R.W., Bruce, N.: Welfare Economics. Basil Blackwell, Oxford (1984)
12. Kneese, A.: Whatever happened to benefit-cost analysis? Water Resour. Updates **116**, 58–61 (2000)
13. Reuss, M.: Water resorces, people and issues: Interview with Professor Arthur Mass. Office of History, United States Army Corps of Engineers (1989)

14. Greenberg, M.: Environmental Policy Analysis and Practice. Rutgers University Press, New Brunswick (2007)
15. Zerbe Jr, R.O.: The legal foundations of cost-benefit analysis. Charleston Law Rev. **2**, 93–184 (2007)
16. Dobes, L.: A century of Australian cost-benefit analysis: Lessons from the past and the present. Working paper 2008–01, office of best practice regulation. Department of Finance and Deregulation (2008)
17. Hanley, N., Spash, C.L.: Cost-Benefit Analysis and the Environment. Edward Elgar, Cheltenham (1993)
18. Russell, R., Baumann, D.: The Evolution of Water Resource Planning and Decision Making. Edward Elgar, Cheltenham (2009)
19. Boland, J., Flores, N., Howe, C.: The theory and practice of benefit-cost analysis. In: Russell, R., Baumann, D. (eds.) The evolution of water resource planning and decision making, pp. 82–135. Edward Elgar, Cheltenham (2009)
20. Buchanan, J.M.: Positive economics, welfare economics, and political economy. J. Law Econ. **2**, 124–138 (1959)
21. Ljusnans vattenregleringsföretag: (2008)
22. Vattenkraft.info: Dönje. In Swedish (2012)
23. Johansson, P.O., Kriström, B.: The economics of evaluating water projects. Hydroelectricity versus other uses. Springer Verlag, Heidelberg (2012)

Chapter 2
The Dönje Hydropower Scenario

In this chapter we present the cost-benefit analysis of the Dönje hydropower plant re-regulation. In Sect. 2.1 the theoretical cost-benefit rule underlying the empirical study is briefly discussed. It is seemingly simple but is derived in [1] for a quite complex open economy setting. We then go on to the empirical study. In Sects. 2.2 and 2.3 the major benefits of the re-regulation are presented. These are environmental such as more salmon in the river, and hence improved fishery, improved aesthetical values, improved canoing and other such leisure activities. A web-based questionnaire was used to assess the willingness-to-pay (WTP) for these benefits. The valuation question is novel in the sense that it introduces intervals (upper and lower bounds for WTP) that are selected by the respondent. The chapter then goes on in Sects. 2.4 and 2.5 to estimate the costs of the proposal. These are mainly the loss of profits by the hydropower company and externalities of coal-fired replacement power. The intricate question how to discount benefits and costs is addressed in Sect. 2.6. The results are summed up and point estimates are presented in Sect. 2.7. A sensitivity analysis of the result of the cost-benefit analysis ends the chapter.

2.1 The Basic Cost-Benefit Rule

In this section we discuss how to design a cost-benefit rule to be used to assess the re-regulations, i.e. changes in water use, suggested by the two considered scenarios. We present a simple general equilibrium cost-benefit rule for a tax-distorted economy. The small or marginal project under consideration diverts water from electricity production to more environmentally friendly uses. The items in the associated cost-benefit rule are briefly discussed here. More detailed derivations are available in [1]; a sketch of the general principles can be found in Appendix A. It should be added that our evaluation is *ex ante*, i.e. we consider re-regulations that have not yet occurred;

Thanks to Springer for allowing us to draw on [1].

P.-O. Johansson and B. Kriström, *Evaluating Water Projects*,
SpringerBriefs in Economics, DOI: 10.1007/978-3-642-36790-8_2,

for a recent *ex post* analysis of dam relicensing in Michigan the reader is referred to [2].

The special feature of the project under consideration is that it involves a private, reasonably profit-maximizing multinational firm in part owned by foreigners. At first glance, the scenario seems inexpensive to Swedes, because a large fraction of the shareholders are not part of the Swedish society. A significant fraction of the loss of profit is borne by individuals outside the conventional definition of a society, unless Swedes have some altruistic reasons to include the well-being of these foreigners in their utility functions. However, if we respect property rights, there is no obvious way to force the firm to deviate from its profit maximizing use of its water use rights. Therefore, the way to proceed is to provide the firm with an incentive to use less water for electricity generation. In effect, this is equivalent to buying back some of the company's water use rights or to in some other way "bribe" the company to change its level of production.[1] The approach means that the assumed *counterfactual* or *baseline* is "doing nothing" or "business as usual". Thus in the absence of the considered project the current regulation is assumed to remain unchanged over the considered time horizon.

In Appendix A we present a simple general equilibrium model of a small open economy. Cost-benefit rules are generated by marginally changing a parameter and tracing the changes from one general equilibrium to another. The parameter used here is interpreted as a contract according to which the firm receives a sum of money in return for a reduction in its electricity generation at the Dönje plant. The resulting societal cost-benefit rules for a *small* change, which is derived in detail for a small open economy in [1], can be stated as follows

$$\Delta W^M = \sum_h WTP_h \cdot \Delta E + \Delta\pi - \sum_h WTA_h \cdot \Delta EM + \Delta T^d \tag{2.1}$$

where ΔW^M denotes the change in societal welfare converted to monetary units, WTP_h is the present value willingness-to-pay (WTP) of individual h ($h = 1, \ldots, H$) for environmental gains, denoted ΔE, associated with the project, $\Delta\pi$ is the present value loss of before-tax profits of the hydropower firm, WTA_h (≥ 0) is the smallest compensation needed to willingly accept increased emissions from replacement power, ΔEM (≥ 0) is the magnitude of harmful emissions emitted by replacement power, and ΔT^d (≤ 0) is a term reflecting the distortion in the electricity market (difference between consumer and producer prices) and some other minor items as explained in [1]. We define a small or marginal project as a parametric change that can be captured reasonably well by a linear approximation (in our case of a social welfare function). This means that second order and higher order terms, denoted R in Eq. (A.4) in Appendix A, are assumed to be approximately equal to zero.

---

[1] From a legal point of view the Swedish water use rights concept is quite complex. In this Brief we will speak of buying/selling such rights. Thereby we simply mean a contract between two parties stating the present value sum of money paid in exchange for a specified change in water use.

The term ΔE covers two distinct environmental consequences associated with diverting water from electricity generation. The first consequence relates to a possibly smoother downstream water flow. There are stochastic short-run variations in supply (a nuclear power plant, for example, might suddenly shut down) and demand (due to a sudden change in temperature, for example) and also forecast errors by producers. Such variations are largely handled by hydropower in the Nordic countries; the reader is referred to [3, 4] for detailed analyses of the properties of different regulating services. However, the other side of the coin is that sudden changes in the water flow may cause damages to the river basin and sometimes creates a moon-like landscape. If the considered change at Dönje causes a less volatile or more smooth water flow, there is a benefit. However, it was deemed likely that the impact is so marginal that the associated willingness-to-pay is set equal to zero.

The second consequence relates to the main purpose of the considered project. This purpose is to improve the recreational and other values of the basin downstream the Dönje power plant. The associated willingness to pay (WTP) for these values is captured by the term WTP_h in Eq. (2.1). Since Fortum has acquired the water use rights it is legitimate to use a WTP concept rather than a willingness to accept compensation (WTA) concept.[2] In other words, the local residents must pay to obtain an improved environmental quality.

In the empirical study it is assumed that $WTP_i \geq 0$, where a subscript i refers to an individual living in the municipality of Bollnäs.[3] Since the scenario analyzed in this study is small we set $WTP_j = 0$ for all individuals j living outside the considered municipality. Thus in Eq. (2.1) $\Sigma_h WTP_h \cdot \Delta E$ is simply equal to the sum of $WTP_i \cdot \Delta E$ for all i, i.e. all residents of the municipality of Bollnäs.

The second term in Eq. (2.1) captures the loss of present value pre-tax profits of the hydropower plant as water is diverted from electricity generation. The single period loss is $\Delta\pi_t = (p_t^s - MC_t) \cdot \Delta x^s$, where the subscript t refers to period t, p_t^s is the spot price at time t,[4] MC_t is the marginal cost, and $\Delta x^s < 0$ is the constant reduction in electricity generation at the plant. An illustration of the profit loss argument is found in Fig. 2.1 where the (highly simplified) "supply ladder" shifts to the left due to the considered project; it is assumed that the annual loss of electricity is equal to 3.7 GWh just as in our scenario. Cheap hydropower is replaced by electricity generated by a marginal supplier. Hydropower, wind power, and nuclear power plants have low but different marginal costs and are to the left on the ladder while fossil-fueled plants (coal-fired, oil-fired and gas-fired) have high marginal costs and hence are to the right of the ladder. The spot price is "determined" by the marginal supplier.

The principal cost for the project is the difference in costs between the two power sources; the reader might note that the difference between the two dark staples is

---

2 For discussion of these concepts the reader is referred to, for example, [5] or [6].

3 In order to simplify notation, people living in nearby communities are included in the municipality of Bollnäs.

4 The firm might also earn a revenue from providing balance services to the power grid but as explained above these services are probably not affected by the considered marginal shift in production.

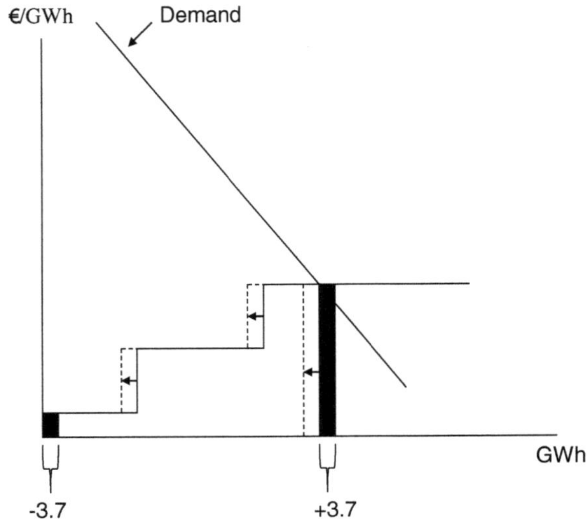

Fig. 2.1 A simplified spot market for electricity. *Source* own work

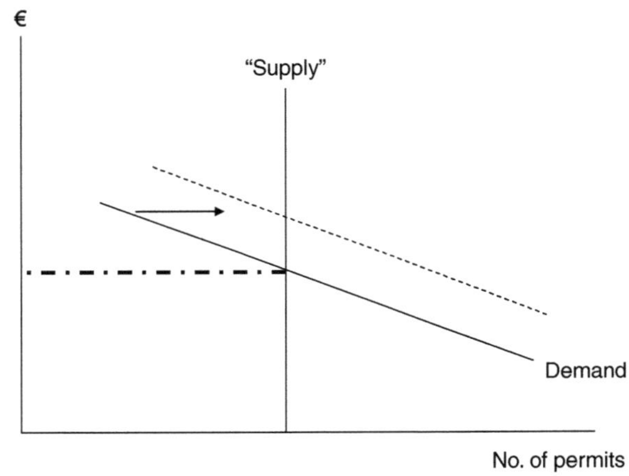

Fig. 2.2 A simple illustration of a permit market. *Source* own work

equal to the annual loss of profits for the Dönje plant if its annual production decreases by 3.7 GWh. In order to arrive at the present value term $\Delta\pi$ in Eq. (2.1) forecasts for p_t^s and MC_t are needed as well as a social discount rate.

In our base scenario a reduction in the electricity generation at Dönje is assumed to be covered by increased production by (foreign) fossil-fired power plants. This will be further clarified when we describe the operation of the Nord Pool spot market. These fossil-fired plants cause emissions of climate gases. However, if there are permit markets net emissions will remain unchanged, as is illustrated in Fig. 2.2. A fossil-fired plant that increases its electricity generation must acquire permits which means that through a price adjustment some other producer will be induced to reduce its emissions, i.e. to sell some of it permits. However, not all climate gases are covered by the European trading scheme. Moreover, emissions of other gases, such as sulphur and nitrogen, might increase as Sweden imports more fossil-based electricity.[5]

A conventional cost-benefit analysis deals with monetary welfare consequences at the national level. The key question is if this implies that consequences of a project occurring outside the borders of the country should be ignored. A standard cost-benefit analysis, just like conventional welfare theory, relies on the concept of consumer sovereignty, so that individual preferences should be respected. Therefore, if Swedes are "nationalistic" egoists in the sense that they care only about damage within the borders of the country, the cost-benefit analysis should ignore any damage caused abroad by replacement power. On the other hand, if Swedes care about the impact of their actions irrespective of where the damage occurs, a cost-benefit analysis should respect this fact. In the base case we will assume that the representative Swede is an altruist in the sense that he or she cares about any damage Swedish actions caused abroad.[6] The term $\sum_h WTA_h \cdot \Delta EM$ is supposed to cover the minimum aggregate (national) compensation needed for accepting the total of all extra emissions, regardless of whether they cause damage domestically or abroad.[7] It is important to realize that we here consider a WTA concept (preceded by a minus sign) rather than a WTP one. The reason is that an increase in emissions causes a loss of welfare.

An alternative to the national level approach used here would be to use a European Union or even a global perspective where the costs to and benefits for the Union or the globe are estimated but such an approach seems to be at odds with the conventional definition of a social cost-benefit analysis which loosely restricts it to dealing with residents of a nation or country. However, there are exceptions, for example, the famous "Stern Review" in [10] of the global costs of climate change and [11] who discuss evaluations at the European Union level. It is also important to stress that this Brief throughout considers small or marginal projects. For recent general equilibrium evaluations of large or non-marginal projects, the reader is referred to [12–14].

Finally, there is the tax term in Eq. (2.1). As is shown in [1] this term is basically the same general equilibrium tax rule as the one stated in Eq. (15) in [15]. In particular, in the present case it reflects the difference between the consumer and producer prices of electricity (and some other, but small, items that are ignored in this Brief). If aggregate demand for electricity is left unchanged by the considered small parametric change, then $\Delta T^d = 0$. This is the assumption maintained in the base case of the study; recall that the loss at Dönje is assumed to be covered by increased production at another power station. On the other hand if the project causes a small price increase some

[5] The power plant acquiring permits must crowd out some other producer demanding permits. From a general equilibrium perspective it is more or less impossible to estimate the net impact on emissions. The outcome depends, for example, on whether a steel producer or a oil refinery is crowded out. Moreover a relocation of plants from the Union to other parts of the world might be "triggered" by a new project requiring permits.

[6] The reader is referred to [7–9] for detailed discussion of different altruism-concepts.

[7] Even a hydropower plant causes emissions in a life cycle perspective. However, since we only consider a marginal change in water use we assume that there is a net increase in emissions if we shift to fossil-fired plants.

consumers might reduce their demand for electricity. In this case $\Delta T^d < 0$ while $\Delta EM = 0$ since, by assumption $\Delta x^d = \Delta x^s$, i.e. no replacement power is needed. This case is briefly considered in the sensitivity analysis.[8]

The cost-benefit rule stated in Eq. (2.1) is remarkably simple. Still and as developed in [1] it resembles the one obtained using a quite complex and tax-distorted general equilibrium model of a small open economy. The framework integrates several key issues, including, but not limited to:

- a contract between the hydropower plant and another party (local residents) generating the general equilibrium cost-benefit rule, which is a cornerstone of our referendum-style Contingent Valuation study
- the tax system in the status quo
- partial foreign ownership of the hydropower company
- trade in electricity, renewable energy (electricity) certificates and carbon emission permits
- externalities of replacement power (generated in other countries)
- value of loss of regulating (balancing power) and other system services
- transmission of electricity modeled as provided by natural monopolies
- downstream hydrological externalities
- environmental benefits (aesthetic and otherwise).

A rather unique aspect of our empirical analysis is that we have assessed the environmental impacts from the perturbation by an actual experiment at the plant. Thus, the perturbations have been implemented in a "test-run" and the ecological consequences thus monitored "live" on-site. We proceed by providing a brief description of the scenarios.

2.2 The Impact of Changed Water Flow at the Hydropower Plant

The natural science team of the research project carried out detailed experiments at the power plant, in order to assess how a changed flow would affect fish ecology. Their study target, Klumpströmmen, is unusual compared to many other rivers and river sections in Sweden. For example, it has not been cleaned for timber floating. Therefore the riverbed is more heterogenous than the average river, even when compared to unregulated rivers. This aspect has a positive influence on potential biological productivity. In addition, Klumpströmmen is an outlet stream and this contributes to productivity, both concerning fish- and benthic fauna. However, the outlet effect is not unique to Klumpströmmen. In fact, many bypass channels are close to dams, which mean that outlet effects should be common in these types of systems.

[8] The reader is referred to [1] for a discussion of how this approach is related to the concepts of the Marginal Cost of Public Funds (MCPF) and the Marginal Excess Burden of Taxes (MEB).

Our scenarios are typical constant flow regimes, with a low winter flow and a higher summer flow. Alternatively, we could have suggested a variable flow to mimic a natural flow regime. However, the power station in Dönje has a maximum capacity which is exceeded at high flows. Therefore, the flow in Klumpströmmen is expected to become variable even in the absence of a legislated two-level flow. The most important aspect that makes our scenario "tick" is that Klumpströmmen is a natural or bypass channel. Before the hydroplant was constructed, Klumpströmmen had about 25 % of the total flow; during extreme highflows excess water can be spilled into the main channel to protect the smaller Klumpströmmen from severe disturbance during extreme high flow events.

Six flow regimes ranging from the present minimum legislated winter flow of $0.25\,m^3s^{-1}$ to pristine low summer flows of $41\,m^3s^{-1}$ were evaluated at the regulated, but uniquely pristine stretch, Klumpströmmen. These scenarios were generated "live", by asking the power company to release certain amounts of water during prescribed periods of time. Thus, in June 2008 (when the minimum required flow is $10\,m^3s^{-1}$), 21 and $41\,m^3s^{-1}$ were released from the plant. In May 2007, during the minimum legislated winter flow, the team looked at flows lower than $10\,m^3s^{-1}$, but higher than $0.25\,m^3s^{-1}$.

Compared to the minimum winter flow, with no suitable habitats for grayling and brown trout, the area of suitable spawning and juvenile habitats at $3\,m^3s^{-1}$ were estimated to 3 ha, and 0.5 for adults. At the present legislated minimum summer flow of $10\,m^3s^{-1}$, the suitable habitats covered an area of about 3 ha for juveniles and 4 ha for adults. At higher flows, these areas increased proportionally less than the corresponding increase in the flow ($21\,m^3s^{-1}$; juveniles 4 ha, adults 5–6 ha and $41\,m^3s^{-1}$; juveniles 6 ha, adults 8–9 ha).

Snorkeling in the river section indicated low relative fish densities, averaging 0.39 grayling per $100\,m^2$, while no brown trout was observed. By electro-fishing, an average of 0.14 brown trout per $100\,m^2$ was estimated, which in relation to reference data indicated a very low density. Based on the collected information, the current flow regulations in the section seem to impair fish populations. With environmentally adapted flows implying c. 1.5 month earlier start and ending of minimum summer flow, and increasing the minimum winter flow from 0.25 to $3\,m^3s^{-1}$, the salmonoid density was predicted to increase by a factor of 3–6 compared to that estimated in the field, see [16, 17]. These predictions, combined with other information about the possible impact of the scenarios, were included in the contingent valuation study, to which we now turn.

2.3 Contingent Valuation Study of Improved Downstream River Basin

The contingent valuation study targeted respondents in the Bollnäs municipality (but the survey also gave some information about WTP in neighboring municipalities). It was preceded by the focus groups analysis described below and the on-site ecological assessment described above. These two studies were basic ingredients in the scenario development. An important step in the development of a survey is to conduct focus groups to test how a general audience interprets the survey (subsequent steps include survey revisions, a pilot study, and a final study. This last step was omitted here, because the web-panel was essentially exhaustive in the Bollnäs area and we did not have resources to undertake additional in-person or telephone interviews). Specifically, our focus groups were designed to ensure, among other things, that the proposed environmental changes and the hypothetical valuation scenario described in the survey were understood by respondents. The reader is referred to [1] for a detailed presentation of the results of the focus-groups study.

Given the insights obtained from the focus-groups, we then proceeded to revise our survey instrument. A web-based sample was obtained providing 136 completed surveys for the Bollnäs municipality, and 200 completed surveys in total. The 64 questionnaires (200−136) not from the Bollnäs municipality came from households in neighboring municipalities (5 cases had invalid zip codes and were deleted from the analysis of residence).

We then undertook a straightforward representativity analysis of the sample. This analysis shows that there are only small differences between our sample and the population of Bollnäs. An interesting fact is that we had a slight underrepresentation of young households and a slight overrepresentation of old, which might not be expected in a web-survey.

The WTP-question was formulated as a referendum in the Bollnäs municipality. This follows a long-standing tradition of local referenda in some countries/states (e.g. Switzerland/California) in general and on water issues in particular. For example, the *Wyoming Preserve Minimum Instream Water Flows Initiative*, May 23, 1982 in Wyoming, USA, mandated leaving a certain level of water in streams so that natural fisheries could thrive. The initiative was set for the 1986 ballot, but did not make it.[9]

In the present study the question was formulated in the following way:

WTP-question introduction *It has become more common in Sweden that those who are affected by local environmental issues are able to vote in local referendums. The following proposal can be viewed as such a local referendum. It is thought to be held among inhabitants of Bollnäs municipality. The purpose of this question is to shed some light on how the average citizen of Bollnäs values a potential change of the water flow in the Bollnäs streams. The change will improve fishing conditions, water ecology and landscape aesthetics. At present, the water rights are owned by the Fortum company. This means that Fortum has the right to produce electricity at the Dönje plant. Suppose that the only possible way to increase the water flow in the Bollnäs streams is to buy back those water rights, by means of a joint action among Bollnäs citizens.*

[9] It was argued that the Wyoming State Legislature in 1985 accomplished substantially the same objectives. Source http://ballotpedia.org/wiki/index.php/Wyoming_Preserve_Minimum_Instream_Flows_Initiative_(1986).

We then described the winter-season proposal.

> ***Proposal to change of the winter season water flow in the Bollnäs streams.*** *The* ***Proposal*** *entails an increase of the winter season water flow from 0.25 to 3* $\mathrm{m^3 s^{-1}}$. *There will be no change in the summer season water flow. This means that the water flow will increase from the power station to Varpen (note: this was described in a map not included here) The proposal is depicted in a series of pictures in the sequel. The total costs for the* ***Proposal*** *is not known with certainty at the present time. Suppose that the referendum is held when the cost has become known. If a majority supports the proposal, it will be undertaken. A "yes" entails each household paying a given sum over a period of 5 years.*

A series of pictures were introduced in order to depict the winter scenario, i.e. SCENARIO 1. For SCENARIO 2, a picture was added showing the summer season change. These and the detailed written information provided about the impacts of the proposed changes can be found in [1].

The next question included an opt-out possibility, i.e. a zero WTP possibility. Those who rejected this option could then state their WTP. If an individual was not able to state WTP as a point he was asked to state it as a freely chosen interval. This last feature is unique, to the best of our knowledge, and opens up for challenging theoretical and econometric exercises.

Turning to the results, it is of interest to consider the "size of the market". Because the scenario entails payment responsibility only for those living in the Bollnäs municipality, we focus on these respondents in the sequel. Around 62 % reports $WTP > 0$ and 38 % reports $WTP = 0$. Thus there is a majority in favor of the project. Further results are summed up in Table 2.1.

The table makes it plain that we do have a sparsity of data, and, again, our analysis here is only to illustrate how to undertake a modern cost-benefit analysis of water use conflicts. In order to obtain average WTP, we simply add the proportion of zero WTP in the data, i.e. those respondents who claim not to be "in-the-market". In so doing, we need to handle the 6 respondents who reported an interval (0,upper). We interpret them to be in the market.

The figure that we use in the empirical CBA is taken from Table 2.1 by including the zero answers. It is calculated as $\frac{36}{135} \times 482,25 + \frac{47}{135} \times 495,08 \approx 300$, where 495.08 is the average of the mid-points. It should be noted that there was no difference in average WTP between SCENARIO 1 and 2. Our hypothesis is that the reason lies

Table 2.1 Descriptives of WTP, conditional on $WTP > 0$

WTP	n	Unique	Mean	0.50	0.75	0.90	0.95
Lower	47	13	214.3	100	100	540	1000
Point	36	13	482.2	250	500	1000	1375
Upper	47	19	775.9	500.0	1100.0	1700.0	2350

Source own work

in the fact that the difference between the status quo and the scenario, i.e. from 10 to $20\,m^3s^{-1}$ in the summer season, is quite small and hard to recognize.

In [1] we present estimates (and computer programs) based on the Jammalamadaka estimator, a Weibull distribution, and the Belyaev-Kriström estimator. These estimates are 286, 258, and 285, respectively, and are hence quite close to the estimate used in the CBA, although we must emphasize that the number of observations is very small. It is noted that [18] report very similar estimates using different estimators but the same data set, their means ranging from 258 to 334, depending on the specific assumptions made. Finally, to repeat, for ease of replication we use the rounded number 300 as our estimate of the average annual 5-year WTP for the project's benefits in the empirical CBA.

2.4 The Cost of Electricity Foregone

The primary cost of the project is the value of the hydroelectricity foregone when water is diverted to other uses. The physical loss (in SCENARIO 1) was estimated to 3.7 GWh per year[10]; see Sect. 3.3 or [1] for the simple engineering formula used for the estimation.

A central part of the CBA is to construct a trajectory for the price of electricity. The point of departure is the spot price at the Nordic (Denmark, Finland, Norway and Sweden) spot market Nord Pool. Typically the marginal supplier on the spot market is a fossil-fired plant (i.e. would be on the final step of the highly simplified supply ladder in Fig. 2.1). This fact means that the spot price will reflect the price of carbon emission permits. Recall that fossil-fired plants must acquire a permit for each ton of carbon dioxide they emit, which implies that their marginal costs reflects the marginal cost of permits. In recent years the price of carbon permits at the Nord Pool market has typically varied between EUR 15 to EUR 20 per EUA, with the holder of one EUA (European Union Allowance) being entitled to emit one ton of carbon dioxide or carbon equivalent greenhouse gas.[11]

In Fig. 2.3 we show how the Nord Pool spot price (system price[12]) for electricity has fluctuated from 1996 to 2010 (inclusive). The "spikes" in 2003 and 2006 are due to dry conditions, leading to a low supply of hydroelectricity, while the spikes in early to late 2010 were due to extremely cold periods and transmission problems. The figure includes the linear trend of the spot price.

There is trade between the Nordic countries and Estonia, Germany, the Netherlands, Poland and Russia. Hence the price on the Nord Pool is not independent of what

[10] In SCENARIO 2 the annual loss is estimated to 14.3 GWh.

[11] See the PDF file entitled "Marknadspriser" at www.svenskenergi.se.

[12] There are sometimes bottlenecks in the transmission of electricity implying that the Swedish price deviates from the system price. Here we ignore such deviations since we focus on long run issues. Historical spot prices are available at Nord Pool's home page: http://www.nordpool.com/asa/. Since inflation has been very low during the considered period we report the spot price in current terms.

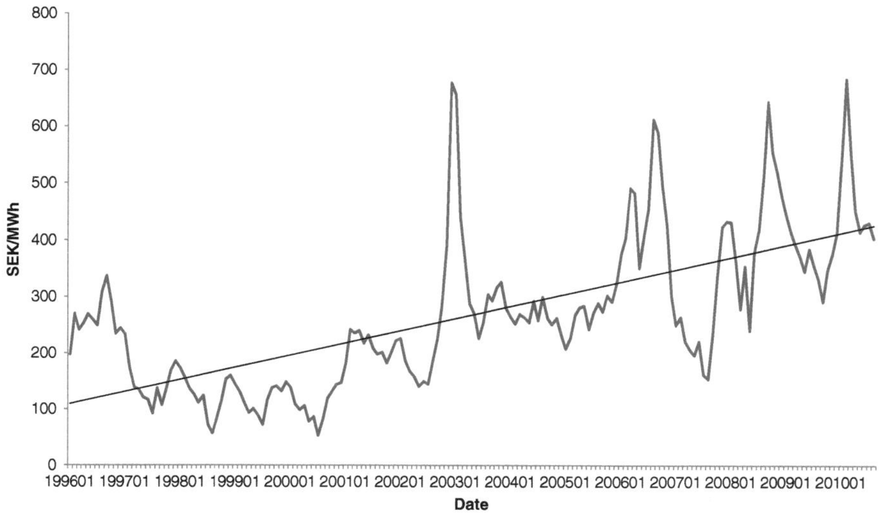

Fig. 2.3 Average monthly spot prices (system price) in SEK/MWh 1996–2010 and the linear price trend on the Nord Pool market. *Source* own work based on data from Nord Pool Market

happens in the electricity markets in the surrounding area.[13] For example, in the dry years of 2003 and 2006, the Nordic countries did import electricity from Germany, Poland and Russia. In particular, the spot market of the Leipzig based European Energy Exchange, EEX, is of interest in the present context. This is because from the point of view of Nordic hydroelectricity producers it seems to be ideal to have unlimited transmission capacity to Germany since the EEX spot price typically is higher than the Nordic one and has a different profile across the day. The German merit order curve looks similar to the Nordic one but there are also important differences. In contrast to the Nordic system, Germany has little hydroelectricity and a lot of lignite-fired power plants. Germany also has pumped-storage plants for electricity generation during peak hours.[14]

In the CBA two different price trajectories are used. The first price trajectory is a conservative one which is aimed at providing a reasonable *lower bound* for the loss of revenues of the Dönje plant. According to this scenario, the price received by Dönje "today" is equal to our estimate of Nord Pool's (system) spot price for an average or "normal" year with respect to precipitation. This price, assumed to be constant over time, is set to SEK[15] 350/MWh (about EUR 35/MWh) and corresponds roughly to the average spot price during the period 2003–2010; the average is SEK 361/MWh.

[13] For a good treatment of European electricity markets the reader is referred to [19].

[14] A pumped-storage plant (typically) works as follows. At night, when electric demand is low, the plant's reversible turbines pump water (uphill) to a reservoir. During the day, when electric demand is high, the reservoir releases water through the turbines.

[15] A Swedish krona (SEK) is here assumed to be worth about EUR 0.1 in the long run, i.e. EUR 1 is equal to SEK 10.

This forecast is consistent with the idea of mean-reversion, which is the tendency for a stochastic process to return over time to a long-run average value. The reader is referred to [20] for further discussion of the concept of mean-reversion in the context of electricity markets.

The second price trajectory is supposed to provide a reasonable *upper bound* for the loss of revenues for the Dönje plant. The best scenario of Swedish hydropower proponents is arguably a "merger" of Nord Pool and the German EEX market. In this scenario Swedish hydropower is sold at the German peak load price from 2030 and on. We assume that the initial average German peak load price is SEK 700/MWh (about EUR 70/MWh), which corresponds to the average price during the period 2005–2008, and that the average price stays constant at this level. As in the first scenario, the initial price received by the Dönje plant is assumed to be equal the "normal" annual spot price on the Nord Pool market, i.e. SEK 350/MWh, and then linearly approaches the German peak load price in such a way that they become equal in 2030. This corresponds to an annual real price increase of 5 % during the period 2010–2030 so that the real spot price reaches SEK 700/MWh in 2030 and then stays constant.

It might be argued that the first price trajectory represents "business as usual" in the sense that there are no major changes in the market. The second trajectory accounts for a possible major long-run structural change in market conditions. In our cost-benefit analyses we will draw on point estimates of the different items. With respect to the price trajectories we assume arbitrarily that the spot price at each point in time is an i.i.d uniform random variable over the interval (p_t^{slb}, p_t^{sub}), where p_t^{slb} (p_t^{sub}) denotes the lower bound (upper bound) estimate of the spot price at time t. This is the simplest possible continuous random process and means that the point estimate of the spot price at time t is equal to $(p_t^{slb} + p_t^{sub})/2$. This means that the initial assumed spot price is SEK 350/MWh and that the price increases linearly to reach SEK 630/MWh after 20 years and then stays constant in real terms. This trajectory yields the same present value revenue loss as if we use what might be termed a *certainty equivalent price* of just above SEK 480/MWh if the discount rate is 3 %,[16] i.e. the same present value revenue loss as if the price is constant over time and equal to SEK 480/MWh. As a comparison it might be noted that the Swedish Energy Agency, which is a government agency for national energy policy issues, in a recent long-run forecast (see [21]), estimates the spot price (or rather Swedish system price) to be SEK 500–520/MWh over the next 20 years in 2010 prices. In contrast to one of our price trajectories this last forecast seemingly ignores any major changes in international transmission lines.

Then we are equipped with the spot price p_t^s and the annual loss of electricity Δx^s in the annual profits loss expression $\Delta\pi_t = (p_t^s - MC_t) \cdot \Delta x^s$. The marginal cost MC_t for Swedish hydropower plants is considered to be low. In fact, according to several studies the variable costs of *new* plants are virtually zero. Here it is arbitrarily set equal to SEK 30/MWh for the existing plant and to zero once it is replaced by new equipment.

---

16 Unless otherwise stated we assume that society is risk neutral.

2.5 Pollution Externalities of Replacement Power

If the electricity foregone at Dönje is replaced it is most likely replaced by Danish coal-based electricity. This is often the marginal source on the Nord Pool spot price market (although supply and demand for electricity shifts over the hour, the day, the season, and so on). It is well-known that coal-fired plants cause emissions that might hurt the health of human beings as well as affect the environment (fauna and flora).

In order to put a price tag on these externalities we will use a simple approach. The estimate is based on EcoSenseLE V1.3, a web-based tool for estimating externality costs within the EU.[17] It provides a rough estimate of the shadow prices of increased morbidity and mortality, damage on crops and materials, and climate gases that are not covered by the European carbon trading system. We include emissions of NO_x, SO_2, particulates, NMVOC, CH_4, and N_2O in the way detailed in [1]. The annual cost is estimated to be EUR 3 039 per GWh (about SEK 30 500). This estimate is assumed to reflect the annual compensation Swedes in total request in order to be indifferent to the emissions per GWh caused by replacement power. Thus in the base case we assume that Swedes are altruists in the sense that they care about damage inflicted on others even if these people, animals, plants, and so on, live in foreign countries.

If an emission is subject to an emission tax the externality is partly or wholly internalized. Since EcoSenseLE does not account for this phenomenon our approach could be interpreted as equivalent to assuming that Swedes are paternalistic altruists. That is, Swedes care about the "physical" impact of emissions ignoring that externalities might be fully or partially internalized through taxation. Even if we accept the altruism-assumption it is far from obvious whether our approach overestimates or underestimates the true damage. On the one hand, some particulates are not accounted for by the EcoSenseLE V1.3 model. On the other hand, replacement power production crowds out some other activity in the carbon permits market; recall that there is a *fixed* number of permits. In principle one would need a *global* computable general equilibrium model linked to an emissions model to be able to track down the net impact on different emission sources. Unfortunately, no such model is currently available.

2.6 Social Discount Rate

There is a huge literature on how to define and estimate a social discount rate. There is certainly no universal consensus with respect to what a social discount rate reflects, its magnitude or even sign. A good overview of different approaches is found in [22]. We will not attempt to summarize the different approaches and views here. We just note that the UK uses a base rate of 3.5 % (including 1 % reflecting reflecting catastrophe

[17] See http://ecoweb.ier.uni-stuttgart.de:80/ecosense_web/ecosensele_web/frame.php. This web-based tool can be used to replicate the externality costs presented here.

risk), Germany's rate is 3 % while France uses a rate of 4 %; see [23]. Reference [24] argues for a standard benchmark European discount rate of around 3–4 % based on social time preference. The European Commission's Directorate General Regional Policy has suggested a 3.5 % social discount rate for most member states including Sweden when evaluating infrastructure investments; see the Commission's guide to cost-benefit analysis; [23].

Both [22] and the Commission's (2008) CBA guide provide estimates of the Swedish social discount rate. The model used by [22] is the one suggested by [25]:

$$r = (1 + \dot{p})^{1-\alpha}(1 + \dot{y})^{\sigma}(1 + i) - 1 \tag{2.2}$$

where r is the discount rate, $\dot{p}$ is population growth, α gives the weight of population size on social utility, $\dot{y}$ is per capita income growth, σ is the coefficient of relative risk aversion, and i is the pure time preference. According to this model one arrives at discount rates of 2.9–3.4 % if the annual growth rate of per capita income is about 1.8 % as was the case for the period 1970–2008; the higher (lower) rate is obtained if the coefficient of relative risk aversion (σ) is 1.26 (1 as in their base case[18]). It might be noted that the result that $\sigma = 1.26$ was obtained as a best estimate by [26] in a cross-sectional study covering over 50 countries and time periods; the reported estimates are seemingly very robust and vary from roughly 1.2 to 1.35.

The model of the European Commission (it's Directorate General Regional Policy, see [23]), is as follows:

$$r = \dot{c} \times em + i \tag{2.3}$$

where $\dot{c}$ is the growth rate of consumption (i.e. $(dc/dt)/c$), em is the elasticity of marginal utility with respect to consumption,[19] and i is the pure time preference. Using this model and assuming that the annual growth rate of Swedish consumption is 1.7 %, one arrives at a discount rate of 3.1 %.[20] It should be mentioned that the social discount rate defined in Eq. (2.3) is a general equilibrium rate corresponding to the classic infinite horizon Ramsey model that is analyzed in all advanced macroeconomics courses and nowadays in an augmented version in a typical advanced course in environmental economics. Maximizing a present value Hamiltonian yields first-order conditions that can be rearranged to yield the rate as defined in (2.3) which then equals the marginal product of real capital. The reader is referred to Eq. (7') on

---

[18] Using Eq. (8) and Table 7 in [22] $r = 100 \times (1.0024^{0.5} \times 1.018^{\sigma} \times 1.01 - 1)$, where σ is either 1.26 or 1. The estimate is based on GDP per capita rather than income per capita (which was not available for the considered time period). See Statistics Sweden: "National Accounts, quarterly and preliminary annual calculations".

[19] From a technical point of view *em* equals a coefficient of relative risk aversion. However, in [23] it seems as if *em* and $\dot{c}$ refer to public rather than private consumption.

[20] Using data stated in Table B.2 on page 207 in [23] but with $\dot{c} = 1.7$ (instead of $\dot{c} = 2.5$) one obtains $r = 1.7 \times 1.2 + 1.1 = 3.1$. The annual growth in real (public as well as private) consumption was around 1.7 % during the period 1970–2006 (and no data are currently available beyond 2006). See Statistics Sweden: "Detailed annual national accounts 1993–2006, some series from 1950 and 1980 (Corr. 2009-02-24)".

p. 40 in [27], with the rate of population growth set equal to zero, or Eqs. (2.8) and (2.10) in [28].

Even with moderate constant discount rates, large future damages have almost no effect on current decisions. For this reason it has become quite common to argue in favor of hyperbolic discounting; see, for example, [29, 30]. Such an approach results in a discount rate that is decreasing over time. Thus future generations are attributed lower discount rates than current ones. In the present context this approach of hyperbolic discounting does not seem to be relevant. The reason is the fact that the decision to change the water use is reversible in both directions at any point in time. At least from a theoretical point of view this means that any generation can sell or buy water use rights. Therefore, we will assume a constant discount rate. It does not seem implausible that there is a consensus emerging within EU that infrastructure project should be assessed using a discount rate of around 3–4 %; the rates mentioned above suggest such a possibility. It would greatly simplify comparisons of different infrastructure investment evaluations if they used the same discount rate in their base case evaluations. This does not prevent authors from strongly arguing for and applying different discount rates in their sensitivity analysis. In any case, we will set the social discount rate equal to 3 %; this rate is in accordance with the Swedish estimates presented above.

In the (stochastic) sensitivity analysis the discount rate will be halved and doubled, i.e. changed to 1.5 and 6 %, respectively, in order to illustrate the sensitivity of the results for the choice of discount rate. It might be noted that the lower rate is close to the one proposed by [10] in his famous "Stern Review" of the costs of climate change while the higher rate is the one proposed by [31], one of Stern's more prominent critics among economists. However, it should be stressed that they evaluate large irreversible changes that might be associated with catastrophic consequences, i.e. "projects" that are quite different from the marginal ones under consideration here. In other words, it is far from self-evident that the rates used in evaluating global climate change are relevant for our scenarios.

2.7 Results

We are now ready to collect the different items to arrive at the outcome of the cost-benefit analysis of SCENARIO 1 (SCENARIO 2 is suppressed in this Brief since from a theoretical point of view there is no principal difference between the scenarios).

The point estimate of the willingness to pay for the scenario is about SEK 300 per household per year during 5 years. This is an average across the two scenarios since we are unable to detect any significant differences between the two scenarios; recall that the contingent valuation study is a small pilot study. There are about 13 000 households in the municipality of Bollnäs. Therefore the point estimate of the aggregate present value willingness to pay of those living today amounts to SEK 18 m given a social discount rate of 3 %. It might be objected that all other things equal we *overestimate* benefits by using the highest willingness to pay estimate among our

Table 2.2 Illustration of a (point estimate) cost-benefit analysis of SCENARIO 1; million SEK

Item	Point estimate
$\sum_h WTP_h \cdot \Delta E$	26
$\Delta\pi$	−56
$-\sum_h WTA_h \cdot \Delta EM$	−2
ΔT	0
$\sum$	−32

Source own work

own estimates. However, we use this approach as a simple way of accounting for the fact that there might be a small willingness to pay also in neighboring municipalities.

We find it unlikely that there is a willingness to pay among those living in other parts of the country. After all, we consider a small project that is not associated with or affect any unique environmental values.[21] Therefore, the "local" present value willingness to pay, i.e. SEK 18 m, is assumed to coincide with the national one for the project under analysis. It might be added that if people living outside the local community are pure altruists, then their willingness-to-pay for the proposal is zero. This is so because in the contingent valuation experiment local residents pay according to their willingness-to pay, i.e. their utility levels remain unchanged. This result is proven by [8].

Future generations will also value the benefits generated by the considered project. If expected life time is 80 years, $1/80$ generations or just over 160 households enter each year, assuming a constant population over time. If their WTP coincide with the one of those living today a present value of over SEK 7 m is added to the benefits[22] so that WTP^{RB}_{NPV} amounts to SEK 26 m ($18.2 + 7.5 \approx 26$) in Table 2.2. It should be noted that in principle the considered scenario is reversible in the sense that one generation might buy the water rights while another might sell them back in order to be able to hire more teachers, more nurses to the elderly care, and so on.

The effects of SCENARIO 1 on short-term (say hourly) variations in the water flow are probably very small. Therefore, our point estimate of the associated WTP is zero.

The point estimate of the present value revenue loss of the hydroelectricity plant is SEK 59 m. We estimate that present value variable cost savings amount to about SEK 3 m. Thus the present value profits loss, denoted $\Delta\pi$ in Table 2.2, is SEK 56 m.

Finally, replacement power is likely to cause harmful emissions of some kinds of climate gases (i.e. gases that are not in the European carbon trading system), sulphur dioxide, nitrogen dioxide, and so on. Recall that the marginal electricity generator in the Nordic market typically is a fossil fuel-fired plant. These emissions cause a negative externality (unless they are optimally taxed), implying that those affected would need a compensation in order to be as well off as without the project. This

[21] If people are willing to pay for virtually all environmental projects it might be reasonable to look for the most cost-effective way of achieving similar benefits to those provided by the scenarios considered here.

[22] $18.2(1/80)32.96 \approx 7.5$, where 18.2 is the WTP of those living today and 32.96 is the present value of a SEK per year for 150 years which is the assumed time horizon.

compensation represents a cost for the project, i.e. the present value compensation, denoted $\sum_h WTA_h \Delta EM$ in Table 2.2, shows up with a minus sign. In the base case we assume that Swedes take full responsibility for any damage their actions cause (in this case through emissions from fossil-fired plants that replace the production loss at Dönje). Thus it is assumed that damage outside the country's borders is valued in the same way as domestic damage.

The estimate is based on EcoSenseLE V1.3, a web-based tool for estimating externality costs within the EU. The annual cost is estimated to EUR 3 039/GWh (about SEK 30 500). This yields a present value cost equal to SEK 4 m if the emissions continue "forever". However, it seems likely that emission restrictions will become tighter over time. We account for this fact by assuming that emissions fade away in a linear way to vanish after 20 years. Therefore the point estimate of the externality cost is SEK 2 m. Since this estimate covers some but not all types of emissions it underestimates the true cost (ceteris paribus). On the other hand, it overestimates the true cost if in the "German" price forecast scenario coal-fired plants are not replacing the production loss at Dönje. For example, the replacement power might instead be provided by pumped-storage plants. The same outcome occurs if the plants crowded out by the Danish plant in the market for carbon emission permits reduce their emissions of other harmful substances. However, there is no European or Global computable general equilibrium model linking production functions to "emission functions". Therefore we are unable to estimate the impact of the scenario on *global* general equilibrium emissions of different particulates. In the sensitivity analysis we will try to account for this uncertainty with respect to the magnitude of the emissions externality.

Since we assume that demand for electricity remains unchanged the tax term ΔT is equal to zero. Adding the different items in Table 2.2 we arrive a a point estimate of the social profitability of SCENARIO 1. According to this estimate SCENARIO 1 causes a loss to society of about SEK 30 m.

2.8 Sensitivity Analysis

A sensitivity analysis shows if the results are sensitive to substantial but plausible variations in crucial parameters. Hence it judges the robustness of the conclusions of the CBA. A one-way sensitivity analysis varies one parameter at a time. If the best estimate is used in the base case CBA, we might be able to locate extreme values in a plausible range of the parameter and use these in the sensitivity analysis. Alternatively, it might be possible to construct a confidence interval (say 95 %) for the parameter. If one is unable to find a value/range for the parameter one could perform a threshold analysis. In such an analysis the parameter is assigned a value such that the outcome of the CBA takes on a chosen value, for example, shows a zero result. If multiple univariate sensitivity analyses are undertaken they are often presented in a

Tornado diagram.[23] In such a diagram the parameters are ordered according to their impact on the result from widest range to most narrow range.

Sometimes one must undertake multivariate sensitivity analyses. For example, parameters might be correlated so that their total impact is larger or smaller than the sum of their impact according to univariate analyses. In the simplest case, one considers 2-way sensitivity analysis where two parameters are varied at a time. A typical approach is to construct a diagram in which the two parameters are varied so as to keep the result unchanged, i.e. one construct a type of indifference curves or isoquants where each curve keeps the result at a particular level. However, it is unlikely that both parameters take on their extreme values together. Therefore, one might try to find a set of parameter values that provide a likely upper bound and a lower bound, respectively, for the outcome.

There are also probabilistic sensitivity analyses where probability distributions are assumed for different parameters. Due to considerations of space we will here undertake a sensitivity analysis with respect to the tax term ΔT. Moreover we will present a probabilistic Monte Carlo simulation (with respect to other critical parameters than taxes) to shed some light on the robustness of the point estimate result presented in the previous section. Further approaches are reported in [1].

2.8.1 Demand Changes

Demand for electricity might change due to small price changes caused by the considered project. Even if such demand changes might seem unlikely they cannot be ruled out. For example, with demand functions generating realistic price elasticities, say -0.3 to -0.4, a price increase of no more than around SEK 0.1/MWh is sufficient to decrease demand by 3.7 GWh. Even if we allocate the entire demand reduction to the household sector, which accounts for 20–25 % of the total electricity consumption, a price increase of around SEK 0.4/MWh seems to be sufficient to accomplish the considered demand reduction (assuming a price elasticity of -0.3). This corresponds to EUR 0.00004/kWh, a quite small number (and of similar size in terms of USD).

Therefore, we supply a rough estimate of the magnitude of the value of such changes. For computational simplicity assuming that only households living in flats (in some parts of the country) reduce their consumption we arrive at the following estimate of the present value loss of governmental tax revenue,

$$\Delta T = \Delta x^d \int_0^{150} [282 + [480 + 282 + 50 + 200] \times 0.25] e^{-rt} \mathrm{d}t, \qquad (2.4)$$

[23] For an illustration, see http://www.tushar-mehta.com/excel/software/tornado/decopiled_help/tornado.htm.

where $\Delta x^d < 0$ is the reduction in final demand for electricity caused by the considered project, the unit tax on households' electricity consumption is SEK 282/MWh as of fall 2010, the spot price is assumed to be SEK 480, the before VAT price a final consumer pays for energy certificates is assumed to be SEK 50/MWh, the variable transmission price (including any cost for regulating services) is assumed to be SEK 200/MWh, and VAT is currently 25 %. Vattenfall, E.ON and Fortum, the three largest players on the Swedish electricity market, have variable transmission prices of SEK 174–240/MWh for (some) consumers living in flats[24] (as of fall 2010). Here we assume that such consumers are those who following small changes in prices adjust their consumption and that the average consumer faces a variable transmission price of SEK 200/MWh. According to Eq. (2.4) we assume that all price and tax components except the spot price remain constant over time. This assumption probably causes an underestimation of the true loss since one would expect at least the unit tax on electricity to be increased over time.

This tax term has a huge impact on the outcome of the CBA. In the considered scenario the social *gross* loss increases by some SEK 65 m. This is due to the fact that electricity consumption in Sweden is subject to very high taxation. The numbers illustrate how the outcome is changed if electricity demand falls at the same rate as production at the Dönje plant. It should be stressed that in this demand reduction case there is no replacement of the electricity foregone at Dönje. Therefore the (positive) amount $\sum_h WTA_h \cdot \Delta EM$ should be added to the (negative) amount ΔT.

We probably overestimate the tax income loss because consumers might increase demand for other taxed goods than electricity (or possibly spend the same amount of money on electricity as initially, depending on the price elasticity of electricity demand); that's why we above speak of a "social *gross* loss". If the net impact is restricted to the electricity tax in Eq. (2.4) the loss is around SEK 35 m. However there is a demand-depressing loss of income through the compensation paid to the hydropower firm in order to induce it to divert water from electricity generation so these numbers might underestimate the true loss.

2.8.2 A Stochastic Sensitivity Analysis Based on Simulation Techniques

A straightforward approach to shedding some potentially interesting light on the uncertainties is to use what is sometimes called Monte-Carlo methods, or "systematic sensitivity analysis" (see [32]). Thus, rather than using a number of different parameter values (typically representing extreme outcomes), we draw values of key parameters from a distribution and then calculate net present value, given the drawn numbers. By repeating this process a large number of times, we obtain a distribution

---

[24] E.ON Stockholm småförbrukarprislista (SEK 239.2/MWh), Fortum enkeltariff, Stockholm (SEK 194/MWh), Vattenfall Söder enkeltariff E4 (SEK 174/MWh). These prices are available on the home pages of E.ON, Fortum, and Vattenfall, respectively (as of fall 2010).

of net present values, representing how sensitive the project is towards stochastically perturbing the key parameters. To implement the approach we use the formulas stated in Appendix B of [1].[25]

We let the interest rate, the price forecast, household annual WTP and the number of years that the negative externality is "alive", be described by distributions. Specifically, we draw the interest rate from a truncated normal with mean 3 %, standard deviation 1 % and lower (upper) limit 1.5 (6 %). There is no particular reason to use a truncated normal, many other distributions will be useful. We use the truncated distribution to ensure that the random variables are within the stated limits. The price forecast (i.e. the constant real growth rate of the electricity price) is taken from a uniform distribution on the interval [0, 0.05] where the limits correspond to the assumed lower bound and the upper bound for the growth of the spot price. Here again one could explore different distributions, this choice simply reflects that we do not have any particular information that would motivate any other stochastic assumption. We assume that the price grows (with a constant rate) over the first twenty years, after which it is constant.

Before turning to the results, let us comment briefly on the assumptions made regarding WTP and WTA in this simulation. Household average annual WTP is drawn from a triangular distribution with limits SEK (203,399) and mode 300.83. This choice is influenced by the analysis in [18]. We could alternatively have used the estimated Weibull from the self-selected intervals. At any rate, total present value WTP depends in the simulation on two factors, the interest rate and household annual WTP. The longevity of the externality caused by replacement power is either 0, 20 or 150 years (the assumed time horizon of the project) with given probabilities.

With these assumptions, we obtain for each draw a particular value of the present value profit loss, present value aggregated WTP, present value aggregate WTA and hence a net value of the project. Each draw is so to speak "one world", the drawn parameter configuration, in which the whole project lives. Each particular value depends in a complicated way on the stochastic assumptions made on the key parameters and the particular functional forms used. Even so, this is a very simple set-up and there are, indeed, many ways to make a more sophisticated simulation. For example, we can let the initial spot-price on electricity be a random variable, the duration of the project might be stochastic and so on and so forth. In addition, we can use other statistical (and more general) distributions. The basic purpose here is to illustrate in a simple way how uncertainty can be addressed in project evaluation. In this Brief we prefer simple and direct approaches and rather try to make the point that sensitivity analyses should always be undertaken in CBA.

In each given run we compute the social profitability of the project, given the values of four random variables. By repeating this process we thus obtain a distribution of possible outcomes that provides useful information for decision-makers with respect to possibility that the project is profitable. Another illustration is provided by what might be termed *cost-benefit acceptability curves* since they yield the probability

---

[25] To implement this in R, libraries `Ryacas`, `triangle` and `msm` are convenient. These libraries are available for automatic download from CRAN.

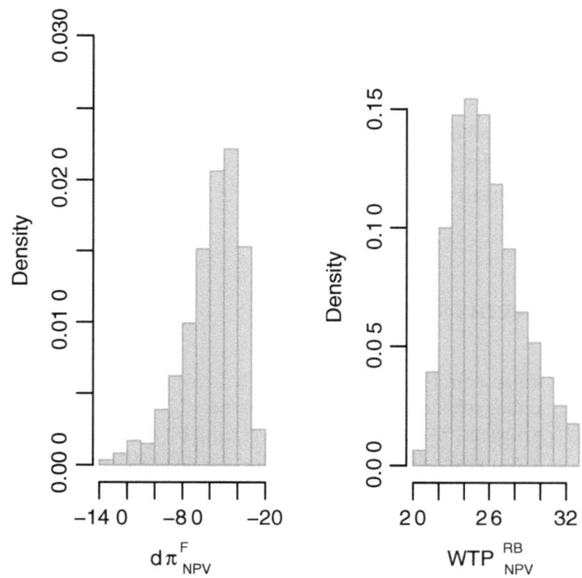

Fig. 2.4 A systematic sensitivity analysis in million SEK based on 10 000 replications. *Source* own work

that the social profitability *exceeds*, say, SEK x million. We believe a decision maker will find such curves more informative and relevant than curves (i.e. cumulative distribution functions) yielding the probability that the outcome is SEK x million or less.

Figures 2.4 and 2.5 summarizes one simulation. Figure 2.4 displays the outcome for the present value of the loss in profit and the present value WTP. There seems to be a slight skewness in the outcomes. In Fig. 2.5 the empirical version of the

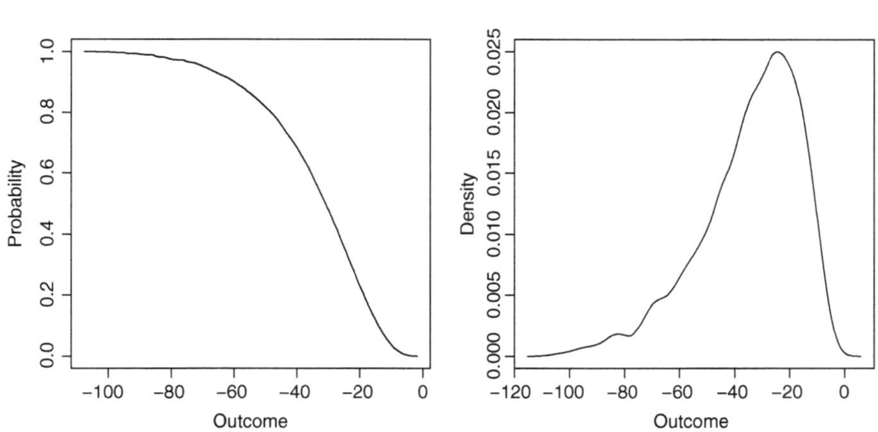

Fig. 2.5 Illustration of a cost-benefit acceptability curve and the associated probability density function from the systematic sensitivity analysis, with outcome in million SEK. *Source* own work

cost-acceptability curve, based on the simulation, is shown. There are several ways to approximate such a curve from data. Here we have simply used the empirical cumulative density functions (using the default settings in the command `ECDF` in R). The right-hand side of this figure uses the default settings of the `density` command in R. Evidently, these simulations suggest that there is very little chance for the project to be socially profitable.

References

1. Johansson, P.O., Kriström, B.: The economics of evaluating water projects. Hydroelectricity versus other uses. Springer, Heidelberg (2012)
2. Kotchen, M.J., Moore, M.R., Lupi, F., Rutherford, E.S.: Environmental constraints on hydropower: An ex post benefit-cost analysis of dam relicensing in Michigan. Land Econ. **82**, 384–403 (2006)
3. Førsund, F.: Energy in a bathtub: Electricity trade between countries with different generation technologies. In: Johansson, P.O., Kriström, B. (eds.) Modern Cost-Benefit Analysis of Hydropower Conflicts, pp. 76–96. Edward Elgar, Cheltenham (2011)
4. Førsund, F., Hjalmarsson, L.: Renewable energy expansion and the value of balance regulation power. In: Johansson, P.O., Kriström, B. (eds.) Modern Cost-Benefit Analysis of Hydropower Conflicts, pp. 97–126. Edward Elgar, Cheltenham (2011)
5. Johansson, P.O.: The Economic Theory and Measurement of Environmental Benefits. Cambridge University Press, Cambridge (1987)
6. Brännlund, R., Kriström, B.: Miljöekonomi, 2nd edn. Studentlitteratur, Lund (2012)
7. Jones-Lee, M.W.: Altruism and the value of other people's safety. J. Risk Uncertainity **4**, 213–219 (1991)
8. Johansson, P.O.: Evaluating Health Risks: An Economic Approach. Cambridge University Press, Cambridge (1995)
9. Kolm, S.C., Ythier, J.M. (eds.): Handbook of the Economics of Giving, Altruism and Reciprocity Foundations, vol. 1. Elsevier, North-Holland (2006)
10. Stern, N.: The Economics of Climate change: The Stern Review. Cambridge University Press, Cambridge (2007)
11. Florio, M. (ed.): Cost-Benefit Analysis and Incentives in Evaluation: The Structural Funds of the European Union. Edward Elgar, UK (2007)
12. Carbone, J.C., Smith, V.K.: Evaluating policy interventions with general equilibrium externalities. J. Public Econ. **92**(5–6), 1254–1274 (2008)
13. Carbone, J.C., Smith, V.K.: Valuing Ecosystem Services in General Equilibrium, WP 15844. National Bureau of Economic Research, Cambridge (2010)
14. Li, C.Z., Löfgren, K.G.: Dynamic cost-benefit analysis of large projects: The role of capital cost. Econ. Lett. **109**, 128–130 (2010)
15. Goulder, L.H., Williams, R.C.: The substantial bias from ignoring general equilibrium effects in estimating excess burden, and a practical solution. J. Polit. Econ. **111**, 898–927 (2003)
16. Rivinoja, P., Gyllenhammar, A., Leonardsson, K.: Predicting European Grayling and Brown Trout densities at various flow scenarios in a regulated section of river Ljusnan, Sweden: Mimeo, Department of Wildlife, Fish and Environmental studies, SE-901 83. Umeå, Sweden (2009)
17. Rivinoja, P., Gyllenhammar, A., Leonardsson, K.: Methods for predicting future salmonid population densities at various flow conditions applied in the river ljusnan. Tech. Rep. 3, Department of Wildlife, Fish and Environmental studies, Swedish University of Agricultural Sciences, Umeå (2010)

18. Yu, J., Ranneby, B.: Estimation of wtp with point and self-selected interval responses. In: Johansson, P.O., Kriström, B. (eds.) Modern Cost-Benefit Analysis of Hydropower Conflicts, pp. 65–75. Edward Elgar, Cheltenham (2011)
19. Roggenkamp, M., Boisseleau, F.: The Regulation of Power Exchanges in Europe. Intersentia, Amsterdam (2005)
20. Johansson, P.O., Kriström, B., Nyström, K.: On the evaluation of infrastructure investments: the case of electricity generation. Invited paper presented at the Eight Milan European Economy Workshop. Published as Working Paper No. 2009–24, Department of Economics and Statistics, University of Milan, Milan, 11–12 June, 2009
21. Statens energimyndighet: Långsiktsprognos 2008. ER 2009: 14 (2009)
22. Valentim, J., Prado, J.M.: Social discount rates. Tech. rep., SSRN.com. http://ssrn.com/abstract=1113323 (2008)
23. European Commission: Guide to Cost-Benefit Analysis of Investment Projects. Tech. rep., DG Regional Policy (2008)
24. Evans, D.J.: Social discount rates for the European union. In: Florio, M. (ed.) Cost-Benefit Analysis and Incentives in Evaluation. Edward Elgar, Cheltenham (2007)
25. Feldstein, M.S.: The derivation of social discount rates. Kyklos **18**, 277–287 (1965)
26. Layard, R., Muyraz, G., Nickell, S.J.: The marginal utility of income. J. Public Econ. **92**, 1846–1857 (2008)
27. Blanchard, O., Fisher, S.: Lectures on Macroeconomics. MIT Press, Cambridge (1996)
28. Barro, R., Sala-I-Martin, X.: Economic Growth, 1st edn. McGraw-Hill, New York (1995)
29. Dasgupta, P., Maskin, E.: Uncertainty and hyperbolic discounting. Am. Econ. Rev. **95**, 1290–1299 (2005)
30. Karp, L.: Global warming and hyperbolic discounting. CUDARE Working Paper 934R, Department of Agricultural and Resource Economics, UCB (2004)
31. Weitzman, M.L.: A review of the Stern review on the economics of climate change. J. Econ. Lit. **45**, 703–724 (2007)
32. Harrison, G., Vinod, H.: The sensitivity analysis of applied general equilibrium models: Completely randomized factorial sampling designs. Rev. Econ. Stat. **74**, 357–362 (1992)

Chapter 3
The −2,+1 Hydropower Scenario

This chapter turns to the Win-Win proposal. The proposal considers a re-regulation of the same river (Ljusnan) as the proposal considered in the previous chapter. The basic idea is to design a package of different measures such that virtually everybody is at least as well off with as without the proposal. We begin in Sect. 3.1 by briefly discussing the theoretical underpinnings of the approach. Then the study is presented. Two dams at the end of the river are removed. This is the −2-part of the proposal. Salmon will then be able to migrate, as noted earlier, some 150 km to the natural barrier. Today they cannot enter the river. There are also other environmental benefits. The company owning the dams is thus far a loser. However the company is allowed to drill a tunnel and install new turbines close to the natural barrier. This which we name the +1-part of the proposal will leave the company a winner from the proposal. A web-based questionnaire was used to shed some light on the views on the proposal by those living in the drainage basin of Ljusnan River. The results of the different parts of the proposal are presented in the chapter.

3.1 A Sketch of the Theory Behind the Approach

We can identify a number of stakeholders that are (or are not) affected by the considered proposal. We take the population to be those currently living in Sweden plus future Swedish generations. One such breakdown of the population is in terms of geographical location although members of a particular group, e.g. stockholders in the company owning the affected hydropower stations, might live in any part of the country or even world since the parent company (Fortum) is a multinational. For simplicity, we here ignore multiple group membership.

1. Shareholders in the hydropower company (Fortum). We restrict attention to stockholders living in Sweden.[1]

[1] For simplicity, we assume that this group is distinct from the other groups identified here.

P.-O. Johansson and B. Kriström, *Evaluating Water Projects*,
SpringerBriefs in Economics, DOI: 10.1007/978-3-642-36790-8_3,

2. Those involved in construction work or dam removals.
3. Those living in the drainage basin
    a. Individuals affected themselves by the environmental and other impacts of the proposal.
    b. Altruists living in the drainage basin that derive non-use values either because they care about others living today and/or are concerned about future generations.
    c. People deriving both use values and non-use values from the proposal.
4. Those living in other parts of the country
    a. Those completely unaffected and ignorant.
    b. Those affected by tax or price or other changes caused by the considered proposal.
    c. Altruists that attribute value to the fact that conditions are improved in the Ljusnan drainage basin.

Obviously this classification is a simplification in that a particular individual might belong to several of these subgroups. Still, it gives an idea of how the proposal under discussion might affect individuals living in different parts of the country.

In Appendix B we provide a more formal presentation of the approach. Just to illustrate the approach, let us consider two groups of stakeholders. The first one is stockholders in the company owning the affected hydropower stations:

$$V^c(p, w^c, \alpha^c \cdot \pi^1 - T^{c^1}) \geq V^c(p, w^c, \alpha^c \cdot \pi^0 - T^{c^0}) \tag{3.1}$$

where a superscript c refers to such a person ($c \in S^c$ and S^c is the set of such individuals living in Sweden), π^1 (π^0) refers to the hydropower firm's present value profits with (without) the project, and T^{c^1} (T^{c^0}) denotes tax payments with (without) the project. Basically the inequality in (3.1) assumes that the firm's present value profits are at least as high with the project as without it, assuming that the individual's tax payments remain unaffected (or decreases). For simplicity, we assume that this individual is not concerned about environmental quality in the considered region of Sweden.

Consider next an individual living in Ljusnan's catchment area. Such an individual is assumed to be at least as well off with the project as without it if it holds that:

$$V^{lj}(p, w^{lj}, -T^{lj^1}, E^1) \geq V^{lj}(p, w^{lj}, -T^{lj^0}, E^0) \tag{3.2}$$

where a superscript lj refers to an individual living along Ljusnan River ($lj \in S^{lj}$). The main impact is through improved environmental quality, i.e. the change in E from E^0 to E^1. During construction of +1 and dam removals, the individual might be affected by disruptions such as noise and heavy transports but such disruptions are assumed to be covered by the E-parameter. One cannot completely rule out that the proposal would affect taxes. In any case, the individual is assumed to make an overall assessment of the proposal's consequences in the questionnaire.

If no individual belonging to these (or other affected) groups is worse off and at least one individual is better off with the considered proposal it satisfies the (strong) *Pareto criterion*. It is important to stress that the assumed *baseline* or *counterfactual* is the *status quo*. In the absence of the proposal, the current regulation is assumed to remain in place for the foreseeable future. This also means that water rights as regulated by Swedish legislation are accepted/respected. It cannot be ruled out that a party would be even better off if only a particular part of the proposal is undertaken than with the entire proposal. For example, shareholders in the hydropower company might be even better off if the company was allowed to install new turbines without having to take down the two dams at the mouth of the river (but some other party might then be worse off than without the proposal).

3.2 Background

The first major commercial Swedish hydroelectric power became operational in 1893 and supplied a mine (Grängesberg mine in the county of Dalarna in mid-Sweden) with electricity. The era of constructing large hydropower plants era ended (as it currently stands) in 1967 with the commissioning of the Sietevare plant (in the river Blaikälven in Northern Sweden). In those roughly 7 decades, hydropower developed rapidly in Sweden. Society's growing demand for energy was already in the early 1900s causing significant discussion, the most prominent example would be the establishment of Great Falls National Park (Stora Sjöfallets nationalpark). The law on national parks was introduced in 1909 and the country's largest waterfall was strongly protected by the formation of Great Falls National Park. The decision was overturned 10 years later when the exploitation of the great northern rivers began. The National Park Act of 1909 was inspired in particular by developments in the U.S. (Yellowstone) and a series of lectures by geographer Hugo Conwentz in 1904 about the benefits of preserving various natural environments, especially older forests. It would take until 1950 before river protection issues obtained the same status.

The many investigations and the decision process that surrounded the development of hydroelectric power in Sweden are too complex and multifaceted to be given a full picture here. However, we note that the battle over Middle Ljusnan, which ultimately was protected, took place in a period of dramatic changes for decision-making processes; several important policy instruments were reviewed (e.g., the Water Act and the national physical planning process) and public resistance from interest groups was mounting. The Water Act of 1918 was revised at the beginning of the 1980s and entered into force in 1983. However, the 1918 Water Act was instrumental in the development of hydroelectric power since it did allow modification of the "natural movement" of running water (according to [1]).

Let us end our brief sketch of the hydro-political discourse by looking towards our neighbors. The history of hydropower in Finland and Norway has many similarities to the Swedish one. Thus there are many examples of far-reaching differences of opinion also in these countries. This applies, for example, to plans to develop the Iijoki River in Finland. These plans were blocked by a strong local opposition, much like the Middle Ljusnan case; see [2] for details. Perhaps the fights in Norway around the Alta River is the quintessential example of how deep a conflict about the management of a river can be. In contrast to the success of conservation interests in Iljoki, the Alta River was exploited and a power plant was operational in May 1987 (according to Wikipedia). A similar event in Sweden is reminiscent of Alta. Sölvbacka Currents in Ljungan's upper part became nationally-known during the 1970s and early 1980s, when "river protectors" tried to stop the water diversion. As a consequence, the Swedish Riksdag, after an initiative from the Government, led by Prime Minister Olof Palme, in a dramatic vote on November 19, 1980 decided to discontinue the exploitation. The state subsequently paid 287 million for the water rights.

3.3 The Proposal and its Main Consequences

The proposal assumes that the last two power plants of the river, i.e. Ljusne Currents and Ljusnefors, are removed. The loss of electricity is compensated by turbines blasted into the rock at the existing power plant Laforsen. There are obvious local consequences of the proposed project when underground turbines are installed at Laforsen and two hydroelectric dams removed at the Ljusnan mouth. The project gives, approximately, in the order of 1000 man-years during the years of work in progress. A schematic diagram of the power plant investment is found in Fig. 1.5. It may be noted that tunnel-based power plants are not entirely uncommon in Sweden and Norway.

The proposal means that power production does not decrease, biodiversity increases, fish migration (salmon) is made possible, implying that natural reproduction is stimulated. Furthermore, the conditions for sport fishing and other social activities are improved. The impact on biodiversity and other characteristics of the ecological system has been investigated by Associate Professor Kjell Leonardsson at SLU, see [3]. The ecological investigations reveal that after removal of the two dams at the mouth of the river it makes sense to build fishways so that fish can migrate up to the natural migration barrier Laforsen. The proposal implies, as explained earlier, that the water flow returns to a more natural state for a substantial portion of the area.

Already back in 1888, i.e. far before the river was developed, the Swedish author, zoologist, and Fisheries Official Rudolf Lundberg, see [4], noted (and we almost quote him but without being able to fully replicate his 19th-century language) that although the greater part of salmon caught in the Ljusnan River was caught in the lower part of the river, salmon penetrated the river all the way up to 153 km from the sea, where the considerable Laforsen prevented further migration. This is the situation that roughly speaking once again will pertain with the considered proposal.

In compact form, the proposal may be summarized as in Table 3.1.

Table 3.1 A compact summary of the effects of the proposal

Measure	Electricity	Energy certificates	Environmental impact	Employment	Direct cost
"+1"	470 GWh	+	+	1,000 man-years	Investment
Ljusnefors	−121 GWh		+	X man-years	VEF + Removal and restoration costs
Ljusne currents	−272 GWh		+	Y man years	VEF + Removal and restoration costs
Fish paths			+	Z man-years	Investment and operating costs
Diverted water for fish migration	≤ −60 GWh		+		VEF
Other			+	?	Reduced costs for dam reinforcements

VEF = Value of Electricity Foregone. *Source* own work

Electricity production in the expanded plant, labeled "+1" in the table, has been calculated using the average water flow at Laforsen which is 149 m^3s^{-1}. For environmental reasons 50 m^3s^{-1} of the flow is assumed to be released into the mainstream river and thus not possible to use for electricity generation. The height is 65 m from Laforsen to the outlet below Forsänget (according to Google Maps). We estimate that this design of the plant at Laforsen provides an additional 470 GWh per year.[2] For Ljusnefors and Ljusne Currents, we use the same type of approach for estimating the loss of power (and necessary data are available at Vattenkraft.info ([5, 6])). Around 120 and 270 GWh respectively are foregone as these two dams are removed. Using the same and well-known way of calculating electricity production throughout contributes to a consistency in the calculations, i.e. if there is a bias it is hopefully similar for all three plants and hence sum to about zero. It might be noted that the consultancy VBB (see [7]) estimates the electricity generation to 400 GWh per year for a similar but not identical underground plant at Laforsen.

As far as we understand, the new power plant at Laforsen is qualified for electricity certificates while the two removed stations do not currently qualify.

In order for fish to be able to migrate up the river water must be diverted from electricity generation at plants downstream "+1", i.e. downstream the new one at Laforsen (in addition to the 50 m^3s^{-1} considered above). In particular, this is true for plants where the current regulation accepts a minimum flow of 0 m^3s^{-1} during some periods. If 5 % of the water at *all* downstream plants is diverted around 55–60 GWh per year is foregone. This is a reasonable upper bound for the cost. We here draw on a scenario developed in [8] where the investment cost of fish paths is estimated to SEK 0.08 billion.

[2] We use the following formula, further explained in [9]: $kWh = e \cdot g \cdot f \cdot h \cdot t$, where the efficiency parameter $e = 0.85$, acceleration due to gravity $g = 9.81 m/s^2$, f is average flow in m^3s^{-1}, h is height in meters, and t is time (= 1 hour). On annual basis $t = 24 \cdot 365$.

Let us now turn to an estimate of the benefits and costs of the proposal for the company owning the affected hydropower plants. The time horizon is set to 50 years which seems reasonable for infrastructure investments (and we employ the simplifying but unrealistic assumption that measures take zero time to implement). The real discount rate is assumed to be 5 % which is the rate for *financial* analysis recommended by the European Commission, see [10]. Electricity certificates are valued to SEK 0.2 per kWh which reflects the average market price during 2011 according to [11]. The spot price is set to SEK 0.45 per kWh, which corresponds to the Swedish Energy Agency's forecast in [12] of average prices in the coming decades. It is quite similar to the assumption employed in the previous chapter.

The present value revenue for the new plant amounts to some SEK 5.6 billion while the loss of profits due to electricity foregone as two dams are removed amounts to around SEK 3 billion if the marginal cost, just as in the preceding chapter, is zero for new plants and SEK 0.03 per kWh for existing ones. The present value loss of profits due to diverted water to enable fish migration is up to SEK 0.45 billion.[3] Add the cost of constructing fish paths and we arrive at around SEK 0.5 billion. The investment at Laforsen is expected to cost around SEK 2 billion. This cost estimate is based on an earlier assessment of a similar underground plant at Laforsen undertaken by a consultancy; see [7]. So far we arrive at a surplus of around SEK 0.1 billion (and up to SEK 0.5 billion depending on the amount of water diverted). However, there are the currently unknown cost items denoted X, Y, and Z in Table 3.1 and the "armament" costs the company operating the plants avoids due to dam removals; one expects heavier flooding due to climate change to cause such extra costs for dam operators. Roughly speaking the proposal seems to be profitable for the company or at least break-even.

We have so far ignored that the operator avoids replacement investments in plants removed in our scenario (and built in 1976 and 1945 with "renovation" 1985, respectively). This provides a reasonable *lower limit* of the considered proposal's profitability. However, sooner or later the dams and turbines and other equipment must be replaced. A reasonable *upper limit* for the considered proposal's profitability is obtained if the two plants would just break-even after replacement investments. Then in economic terms, the annual net contribution is not 470 - 121 - 272 GWh but 470 GWh. This increases the present value of revenues to about SEK 5.6 billion if refurbishment takes place immediately. Thus there is a considerable surplus if costs amounts to up to SEK 2.5 billion (i.e. 2 + 0.5 as above). Even if the discount rate is increased to 10 % the net present value revenue is SEK 3 billion, i.e. exceeds the estimated investment cost plus the cost of diverted water.

The truth probably lies somewhere between these two quite extreme scenarios. Therefore, it is far from unlikely that the hydropower operator can recoup the investment. It should be emphasized that our calculations suggest a more favorable outcome than what is assumed in the web survey presented below, where electricity production is assumed to be unchanged. The reason is that we got access to the

[3] $\int_0^{50}[((0.45+0.2)\cdot 470-(0.45-0.03)\cdot(121+272+58.5))\cdot 10^6]\cdot e^{-0.05\cdot t}dt = 5.6-3-0.45$, where the annual loss of water due to diverted water is 58.5 GWh.

report by [7] and undertook the more detailed calculations after the web survey was launched.

Finally, it should be stressed that we do not try to evaluate social benefits and costs as part of a cost-benefit analysis in this section. Rather we look at the project's possible profitability (before taxes) for the company operating the affected plants. The reason is that the study's aim is to test the hypothesis that everybody, including the (domestic) owners of the company, gains or at least do not lose from the proposal under investigation. However, it should be recognized that the proposal affects a region with quite high unemployment. Therefore, the employment consequences in Table 3.1 are of some interest and importance. In addition to the hydropower company there are groups of employees that reasonably are positively affected by the proposal. Similarly, if the company earns a profit from the proposal it will probably pay more in taxes. If so it will benefit Swedish taxpayers. The table also makes it clear that there are externalities associated with the considered proposal. We now turn to a presentation of our approach for covering the local residents views on the proposal.

3.4 Web Survey

The proposal affects parts of the Ljusnan drainage basin. This basin covers 4 municipalities: Bollnäs (26,000 inhabitants), Härjedalen (10,000), Ljusdal (19,000), and Söderhamn (25,000). Thus in total around 80,000 people might be directly affected by our proposal. In order to shed some light on how the proposal was viewed by those living in the area a web survey was used. The sample was stratified (sex, 18 years or older) and taken from a survey web panel administered by the firm Norstat. Since the panel was considered too small from a statistical point of view, a random sample of additional people was recruited through telephone contact. Once registered for the panel the respondents were drawn randomly from this "enlarged" panel. The survey was conducted in April-May 2010. In total 445 complete questionnaires were obtained and according to Norstat there were 184 non-respondents. A follow-up of those who did not respond in Söderhamn indicates that they do not differ from those who responded.

The web survey instrument had several ingredients in addition to a set of questions:

- A video (optional) around 6 min informing about the proposal and its consequences.
- Written information, about the same as provided in the video (optional).
- The sketch in Fig. 1.5 in Chap. 1 of the new plant at Laforsen (tunnels, turbines, outlet in Ljusnan).
- Photographs showing the status quo at Ljusne Currents and Ljusnefors and the likely situation in the short-run and in the long-run (when trees and other plants have adjusted) after dam removals.

Fig. 3.1 Illustration of current situation (*upper panel*) and expected long run situation after removal of dams (*lower panel*). *Source* Artwork and photos by Kalle Prorok for online survey

In Fig. 3.1 we replicate one of the set of photographs shown to the respondents. The upper panel shows the dams at Ljusne Currents and Ljusnefors. The lower panel illustrates the expected view in the long run after dam removals.

In Appendix C a compact summary of the web-questionnaire is found. In total it contained 14 questions. Before facing the "valuation" question, the respondents were briefed on the considered proposal. This briefing highlighted the following points:

- The last two power plants of the river (Ljusne Currents and Ljusnefors) are removed.
- A tunnel is built from Laforsen to Forsänget and turbines are installed at the existing power plant Laforsen.
- The new tunnel means that the winter flow of water in the Middle Ljusnan roughly corresponds to the flow before the river was regulated.
- Energy production is assumed to be approximately unaffected.
- Fishways etc. are arranged so that salmon can migrate up the river from the sea to Laforsen (the natural migration barrier). Today salmon do not migrate up the river.
- We estimate that the total amount of adult salmon returning annually to their spawning grounds in Ljusnan may amount to approximately 5,000 individuals, with an average weight of about 7-8 kg.
- Approximately 2,500 of the salmon will be able to reach spawning areas downstream Laforsen (while others have spawning areas further down the river as detailed in the questionnaire but suppressed here).

Table 3.2 The "Referendum"

Response	Söderhamn	Ljusdal	Bollnäs	Härjedalen	Total
Definitely an improvement	0.38	0.28	0.33	0.17	0.32
Perhaps an improvement	0.34	0.45	0.40	0.56	0.41
Neither better nor worse	0.07	0.08	0.06	0.10	0.07
Perhaps a worsening	0.03	0.02	0.06	0.00	0.03
Definitely a worsening	0.01	0.02	0.02	0.00	0.01
Don't know	0.18	0.16	0.12	0.17	0.16
Sample proportion	0.37	0.26	0.28	0.09	1.00

Source own work

- The project cost is paid by the hydropower plant owner.
- Construction is expected to generate at least 1,000 man-years and take several years.

An optional more detailed presentation followed. Then the respondent was asked the following question: *What do you think of this proposal (in comparison to today's situation)*?

- Definitely an improvement
- Perhaps an improvement
- Neither better nor worse
- Perhaps a worsening
- Definitely a worsening
- I don't know

Thus there was no willingness-to-pay question, in sharp contrast to the study in the previous chapter. This fact is explained by our attempt to test if the proposal satisfies the Pareto criterion (rather than, say, the Hicks-Kaldor criterion).

A conventional ballot question only contain "Yes" and "No" and the implicit option to abstain from voting. This binary choice question approach gives a clear-cut image of preferences for a project. In contrast to such a conventional referendum, our respondents were allowed to express uncertainty about the proposal. As is seen above the "Yes" and "No"-options are split in two sub-choices and there is an "indifference" option. Moreover, the respondent was not required to take a stand for or against the proposal, so that "I don't know" is a valid answer. This approach gives a more nuanced picture of the attitudes towards the proposal, relative to a binary choice question. In particular, it is reasonable in this type of experiment to emphasize that uncertainty exists. For example, if the project becomes a reality many detailed issues have to be resolved. Our intention was not to provide the respondents with a detailed operational proposal, as such a proposal would be very extensive. Moreover, many scientific and economic parameters are still uncertain.

The results are summarized in Table 3.2. The overall impression is that a large majority is in favor of the proposal. Nearly 75 % consider the proposal either a definite or perhaps an improvement. Härjedalen, where the lowest proportion finds the proposal a definite improvement, is situated at the upper part of the river. Those living there are only marginally affected by the proposal unless they fish or are involved in other leisure activities further down the river (or derive some kind of non-use value from the proposal). The new plant at Laforsen and the river section Middle Ljusnan are situated in the municipality of Ljusdal while Bollnäs and Söderhamn are the two eastern municipalities; Bollnäs city is situated some 40 km from the coast and Söderhamn city is on the coast. There are some differences between these 3 latter municipalities but we are unable to say whether they are due to random or systematic factors. In any case, overall only 4 % explicitly opposes the proposition but just 1 % find it a definite worsening.

There are two groups of people that are slightly difficult to classify. Around 7 % find the proposal to neither improve or worsen conditions. The most obvious interpretation is that these voters are indifferent. They are about as well off with the project as without it. However, one cannot rule out that some pick the alternative because they find the proposal difficult to understand and judge or because they perceive it to be too vague to vote yes or no to.

The second group consists of the 16 % who picked the alternative "I don't know". One reason might be that a person is indifferent and hence state I don't know. It is not entirely obvious why such a respondent did not choose "Neither better nor worse". A possibility is that a person who is uncertain about whether the aggregated impact is positive or negative is more likely to select a don't know option rather than a neither better nor worse option. Some support for this interpretation is provided by the fact that a not uncommon argument for picking the don't know option was that the respondent did not consider himself/herself sufficiently familiar with the issue.

If the 7 % voting for the alternative "Neither better nor worse" are assumed to be indifferent there are in total 80 % who are better off or at least not worse off with the proposal. If those stating "I don't know" are assumed to be indifferent the percentage increases to 96. Regardless of the interpretation an overwhelming majority is in favor of the proposal. Stated in terms of the strong version of the Pareto criterion it is satisfied for between 80 and 96 % of the respondents. To repeat, the strong version of the Pareto criterion recommends a project if some people are better off while nobody is made worse off. If anything and considering the fact that the considered re-regulation relates to hard resource use conflicts our proposal comes close to unanimity.

3.5 A Sum-Up of the Proposal's Impact

In this chapter we have indicated how some different "stakeholders" are affected by a re-regulation of a river. Let us end by summing up the consequences for the more important groups that one can identify in this context.

An attempt to provide a rough indication of the way the proposal impacts different groups of the current generation is provided in Table 3.3. A plus sign is taken to mean a non-negative impact while a minus sign indicates a strictly negative effect. The impact of the proposal on the hydropower company's profits is most likely strictly

Table 3.3 A summary of the proposal's impact

Group	Impact
Company shareholders	+
Construction workers	+
Most River basin residents	+
Small minority of such residents	-
Taxpayers	+
Others	+

Source own work

positive. Thus one expects its shareholders to be better off. Here we restrict attention to domestic shareholders since we stick to those living in the country. The different measures, both construction of the new plant at Laforsen and the removal of the two plants at the river's mouth will provide new employment opportunities, at least for a number of years. Therefore we hypothesize that what we term Construction workers in Table 3.3 will win from the proposal. Most residents in the drainage basin gain according to the web survey. However, there is a small group of such people who feel that things will be worse with the proposal.

Those living in other parts of Sweden might be affected through tax payments. The hydropower company and construction workers are expected to pay more in taxes. Moreover total electricity generation is likely to increase and hence putting a downward pressure on electricity prices. The group that we term Taxpayers in Table 3.3 is therefore expected to win from the proposal. As discussed in Sect. 3.1 some might also be altruists caring for those living in the considered drainage basin (including plants and fish species and so on). Such people are expected to derive non-use values from the proposal since it has a positive impact on (most) people and other species living in the Ljusnan area.

The last item in Table 3.3 is termed Others. What we have in mind here is in particular the fact that electricity generation from a relatively clean source increases. This is expected to "crowd-out" other sources. The marginal plant on the Nordic electricity market is typically fossil-fueled. Therefore one would expect the proposal to reduce harmful emissions to the environment in the way discussed in Sect. 2.5. Even if this reduction happens to appear abroad, Swedes might be altruists/concerned in the way explored in the previous chapter; see Sects. 2.1 and 2.5. If so, the proposal provides still another benefit. This also highlights the close links between the approaches used in the two studies reported in this Brief as well as that both approaches require insights in welfare theory.

Finally, we expect the impact on future generations to be of the same qualitative nature as in Table 3.3. Therefore our conclusion is that the proposal under consideration comes very close to satisfying the strong Pareto criterion. In this sense the approach points at the possibility to design *Win-Win* proposals to handle difficult natural resource conflicts.

A final issue deserves a brief comment. From the point of view of the hydropower company the best option would be to only undertake the "+1" part of the proposal.

However, it might mean a worse situation for those living in the river basin. There is also a psychological dimension. Many of those affected by dams feel that the owners of hydropower stations take all the benefits (possibly abroad) while the local residents pay the environmental and recreational price of hydropower development. There is a quite extensive bitterness and hence hostility toward big hydropower producers. Therefore, one would expect many local residents to oppose a "+1"-proposal. They would see it as a further exploitation of the environment and the local population.

It is difficult to say if the *overall* proposal would pass a simple cost-benefit test if the *counterfactual* is the "+1" part rather than "doing nothing". The central question is how much people are willing to pay for the switch and if this sum of money is sufficient to cover the extra cost (in terms of electricity foregone when the two dams at the end of the river are removed). If the "+1" part is evaluated versus doing nothing the question is if the winner, i.e. the hydropower company, is able to compensate those who perceive that they lose from the project, as indicated above. A Krutilla-Fisher argument, see [13], might indicate that this compensation possibility deteriorates over time. They find it likely that the benefits of development of a natural resource are decreasing over time. One reason being technological change over time. The benefits of preservation, on the other hand, are likely to increase over time. One reason being that technological change can hardly produce close substitutes to environmental resources while increasing incomes tend to increase the willingness-to-pay for environmental resources (if they are normal, as is typically assumed). Reference [14] develops these ideas in detail.

References

1. Jakobsson, E.: Industrialization of rivers: Studies of Swedish hydropower 1910–1918, Department of History. University of Gothenburg, (1996) (In Swedish)
2. Rusanen, J.: Role of the local people in the utilization of water resources: A case study of the river Iijoki in northern Finland. Nordia **23**, 119–208 (1989)
3. Leonardsson, K.: Environmental restorations in hydro power regulated rivers - Where, when, and how can ecological improvements be expected? Working Paper 5. Department of Wildlife, Fish and Environmental Studies, SLU (2010)
4. Lundberg, R.: Meddelanden rörande Sveriges fiskerier (Notices concerning Sweden's fisheries). Wilhelm Bille, Stockholm (1888)
5. Vattenkraft.info: Ljusne strömmar. (2012a) (In Swedish)
6. Vattenkraft.info: Ljusnefors. (2012b) (In Swedish)
7. VBB (Vattenbyggnadsbyrn AB): Upper middle Ljusnan. Environmentally friendly hydroelectric power. Consultant's unpublished report. (2001) (In Swedish)
8. Länsstyrelserna, Gävleborg, Jämtland, Bollnäs, Härjedalens, Ljusdals, och Söderhamns kommuner, O.: Omprövningar av vattendomar inom Ljusnans avrinningsområde: Del 1 huvudrapport. http://www.ljusnanriver.se/rapport_1.htm (2003) (In Swedish)
9. Johansson, P.O., Kriström, B.: The Economics of Evaluating Water Projects: Hydroelectricity Versus Other Uses. Springer Verlag, Heidelberg (2012)
10. European Commission: Guide to cost-benefit analysis of investment projects. Tech. rep, DG Regional Policy (2008)
11. Swedish Power Grid: Cesar elcertifikat. (2012)
12. Swedish Energy Agency: Long-term forecast 2010. ER 2011:03. (2012b) (in Swedish)

13. Krutilla, J.V., Fisher, A.C.: The Economics of Natural Environments. Johns Hopkins University Press, Baltimore (1975)
14. Smith, V.K.: Technical change, relative prices and environmental resource evaluation. Resources for the Future, Washington (1974)

Chapter 4
A Brief Comparison of the Approaches and an Outlook

In this Brief we have assessed two different proposals for re-regulation of a river. One proposal relating to a single hydropower plant has been evaluated in terms of a detailed cost-benefit analysis. The other proposal has been set up such that most, if not all, affected parties are better off or at least don't lose from it.

The general equilibrium cost-benefit rule used in Chap. 2 might seem simple. Still, the framework integrates several key issues, including, but not limited to: a contract between the hydropower plant and another party (local residents) generating the general equilibrium cost-benefit rule; the contract is a corner stone of our referendum-style contingent valuation study; the tax system in the status quo; (partial) foreign ownership of the plant; trade in electricity; trade in renewable energy certificates; trade in carbon emission permits; externalities of replacement power (generated in other countries); value of loss of regulating (balancing power) and other system services; transmission of electricity modeled as provided by natural monopolies; downstream hydrological externalities and environmental benefits (aesthetic and otherwise). This quite complex framework is used to assess a proposal according to which water is diverted from electricity generation to the natural river channel (dryway) creating environmental and recreational benefits.

The approach, further developed in [1], provides a detailed cost-benefit manual that can be used to assess reasonably small re-regulations of rivers; huge re-regulations that significantly alter relative prices probably need further tools such as computable general equilibrium models. But it can also be used to evaluate other development projects relating to natural resources.

The second study reported in this Brief draws on what we term a Win-Win approach. The idea is to design a proposal such that virtually everybody is better off or at least not worse off with the proposal. If successful the proposal passes a (strong) Pareto test. The approach is applied to a hypothetical re-regulation of a Swedish river. Two dams at the mouth of the river are removed and as a compensation the hydropower company is allowed to build an underground plant some 150 km upstream. The approach more than compensates the hydropower company and will, if some further measures are undertaken, allow salmon to migrate to the

P.-O. Johansson and B. Kriström, *Evaluating Water Projects*,
SpringerBriefs in Economics, DOI: 10.1007/978-3-642-36790-8_4,

natural barrier. We test the hypothesis that everybody affected is better off. It should be noted that there is no monetary compensation involved here. Rather the aim is to design the project such that there is a Win-Win situation.

What would be the outcome of a conventional CBA of this "-2,+1" proposal? Since basically everybody gains from the proposal it would pass a simple CBA-test. This would hold also if we apply a social welfare function. Deriving cost-benefit rules using such a function weights each individual according to the properties of the welfare function. In terms of just two individuals a simple rule (expressed in social welfare units) might be:

$$\Delta W = W_1 \cdot \lambda_1 \cdot CV_1 + W_2 \cdot \lambda_2 \cdot CV_1 \tag{4.1}$$

where W_h is the marginal welfare weight attributed to individual h, λ_h is the marginal utility of income of individual h, and CV_h is the willingness-to-pay for the proposal by individual h. If $CV_h \geq 0$ for both individuals and strictly positive for at least one of them, the proposal will increase societal welfare irrespective of the magnitude of the marginal social utilities of income ($W_h \cdot \lambda_h$) of the individuals. From an ethical point of view the test in Eq. 4.1 is more "demanding" than the simple Hicks-Kaldor test. The latter basically recommends a project if the sum of willingnesses to pay is positive, i.e. if those who gain are able, at least hypothetically, to compensate those who lose. The test in Eq. 4.1 produces the same outcome as the Hicks-Kaldor test if society's welfare distribution is optimal (and the project is small) so that $W_h \cdot \lambda_h$ is the same for everybody; recall that then $\Delta W = k \cdot (CV_1 + CV_2)$, where k is the common marginal social utility of income. In any case, the Win-Win approach is independent of the actual distribution and the marginal social utility attributed to different groups in society since the proposal is (assumed to be) designed such that $CV \geq 0$ for everybody.[1]

Still there is a caveat. There might be "partial" re-regulations that are more socially profitable than "-2,+1", at least according to the Hicks-Kaldor criterion. For example, allowing the hydropower company to install new turbines at Laforsen, i.e. the "+1"-part of the proposal, without dam removals *might* (but need not) produce a larger surplus than the entire proposal. However, such a partial approach might create a public outcry and heavy resistance not only from locals but also from influential political and environmental circles. Recall the sharp conflicts on river development issues indicated in Sect. 3.2. A cornerstone of our Win-Win approach is to avoid or at least minimize such resource use conflicts.

It should be mentioned here that the middle section of the river under discussion (Ljusnan) enjoys a very strong protection under Swedish law and our proposal is not directly feasible without legislative change. The current law "prescribes" that each plant is handled/re-regulated in isolation which rules out our "package" of measures. Our starting point is, nonetheless, that all Pareto-sanctioned improvements should

---

[1] Of course there might be those who find even this assumption problematic, for example, those (Rawlsians) who only proposes measures that benefit the poorest irrespective of their impact on other groups.

be implemented, since they imply that at least some consider themselves to be better off without someone else being worse off. Although such changes are extremely rare in practice, we have nevertheless, optimistic (or naive, depending on how you want to look at it), found it interesting to analyze the proposal on this basis.

Our proposal has some similarities to a policy that was applied in California in the early 1970s for air emissions, an idea that has gained ground in other areas, including natural resource policy. The authorities in California were wrestling with the question how to allow new investment without giving up on environmental quality. "Offsets" means that permission for the expansion was conditioned on air quality was not deteriorated; an expanding company was thus able to sign a contract with another company who then committed itself to reducing its emissions accordingly.

This interchangeability idea has since been applied elsewhere. The Australian authorities in the so-called "Bubble Licensing Scheme" permitted transfers among a large number of plants in a defined region. Individual plants can vary phosphorus and nitrogen emissions in the bubble. In Australia, there is also the Bush Broker, whose basic idea is captured by the following quotation: "In most cases the clearing of any native vegetation that requires planning approval must be offset by a gain elsewhere. Offsets are permanently protected and linked to a particular clearing site." Ref. [2], p. 1.

Since 1972, wetland use in the United States has been regulated by the Clean Water Act. The Act of 1972 means that permits were required for a project that caused harmful interference in a wetland. Corps of Engineers permitted projects, which caused harmful interference, provided that project owners could provide a corresponding offset elsewhere. According to [3] the destruction of approximately 10,000 ha of wetland was allowed, but this was offset by the creation, restoration or protection of nearly 17,000 ha of wetlands elsewhere. In Sweden the same substitution idea has been applied, for example, with respect to infrastructure investment.

There are clear connections between these developments and environmental legislation in EU countries with respect to emissions from stationary sources. In broad terms, the policy has evolved towards a more flexible approach to substitutability. The European emissions trading scheme for carbon dioxide allows a plant to increase its emissions by purchasing emission allowances. In turn this means that emissions are reduced elsewhere in Europe. In both an environmental and a legal sense, it makes no difference where the reduction is made. The underlying environmental global nature of the problem, i.e. greenhouse gas concentration in the atmosphere is independent of the emission source's geographical location, makes such a solution quite natural. Environmental problems associated with hydropower are in many ways of a different character and it is a delicate task to weigh the environmental benefits of an investment in river A against an investment that cause environmental damage in river B (or for that matter, to weigh a deterioration towards an improvement in one and the same river). The environmental impacts of actions in rivers are not independent of the location in the same way as in the case of greenhouse gases. Our proposal should still be considered in light of a revised view on the value of flexibility in environmental law around the world.

It will be useful, given the discussion above, to close by relating our approach to a few recent dam removal proposals in the U.S. The Klamath River flows southwest through Oregon and northern California. It is around 420 km long and ends in the Pacific Ocean. There are six hydropower developments on the river (and one plant that is located on Fall Creek, a tributary to the Klamath River). Due to the dams (and to some extent farm production) salmon runs have decreased sharply in the past century—from millions of fish to less than 100,000 in most years.

In 2010 almost 50 entities (stakeholders), including the farmers, the tribes and the hydropower corporation (PacifiCorp) signed the Klamath Hydroelectric Settlement Agreement.[2] This agreement specifies a list of preliminary requirements that must be completed-studies, interagency agreements, and approvals all the way up to Congress and the Interior secretary. The ultimate goal of the agreement is the restoration of the Klamath River Basin including removal of 4 dams (Iron Gate, Copco 2, Copco 1, and John C. Boyle) by 2020, restoration of habitat within the Basin, and negotiation of a water supply schedule among the many competing water users in the Basin. The removal plan has the backing of several Native American tribes on the Klamath who rely on the river for salmon fishing, as well as farmers who depend on its water for irrigation. Removal of the dams would allow the fish to return to their historic and more productive cold-water mountain streams, which are currently blocked by the dams, and help resolve disruptions to the main stem Klamath. For some details on the issue the reader is referred to [4].

The cost is estimated to be around USD 1 billion, of which the hydropower corporation is expected to cover half and the remainder is covered by federal tax money. Possibly PacifiCorp is backing the plan because it expects it to be cheaper than (being forced) to undertake alternative measures that would allow fish passage if the dams remain in operation. The plan is also supported by several Native American tribes on the Klamath who rely on the river for salmon fishing, as well as farmers who depend on its water for irrigation.

A comprehensive choice experiment (stated preference survey) was undertaken by a team of well-known researchers, see [5], to shed light on the willingness-to-pay for the project in three regions (a 12-county around the river, the rest of Oregon and California, and the rest of the country). The survey was administered mainly as a mail survey but the respondents had an option taking the survey via the Web. The total response rate was around 33 %. Around 45 % of respondents living in the 12-county area, 29 % in the rest of Oregon and California, and 34 % in the rest of the US voted No to the action plan. Interestingly (but not easily explained), the hardest resistance is found among those living closest to the Basin. Possibly, many of those living close to the river feel that the company responsible for the current situation, i.e. the hydropower firm, should pay for restoration. It might also boil down to perceived *water rights*, i.e. those living in the affected area feel that they should not pay for restoration, as suggested in the survey question, since they perceive that they have

[2] The agreement as well as much other information is available at http://klamathrestoration.gov/ which is the official website of the Department of the Interior and other agencies involved in carrying out obligations set forth in the Klamath Hydroelectric Settlement Agreement.

the historical right to the water. It must be underscored though that this is our own hypothesis. The reader is referred to Chap. 8 in [5] for a detailed estimation of the willingness-to-pay for the likely impacts of the project among people in different geographical parts of the U.S.

This is undoubtedly a very complex project and the final outcome remains to be seen. It also differs from the "-2,+1"-proposal since many Americans, both among those living in the Basin and those living elsewhere, feel unhappy with (paying for) the project. It is not designed so as to make everybody better off. Rather a conventional cost-benefit analysis seems to be the relevant evaluation tool.

Another example is provided by the Elwha and Glines Canyon Dams on the 72 km long Elwha River in the state of Washington. The dams were built in the early 1900s and they have significantly changed the watershed. In particular, they confined migrating salmon to the lowest 8 km of the river. In 1994, the Department of Interior determined it was necessary to remove both dams in order to restore the river and the fisheries. The federal government purchased the dams from the operator for USD 29.5 million in 2000. The dams remained in operation under the management of the US Bureau of Reclamation while plans for removal were developed. Dam removals began in 2011 and are expected to continue over period of a couple of years. For some details on the case the reader is referred to [4]. The National Oceanic and Atmospheric Administration (NOAA) has initiated a Restoration Valuation Study that is expected to be ready by the end of 2013.[3]

Still another example is the Condit Dam on the White Salmon River, also in the state of Washington. It was completed in 1913 and was located about 5 km upstream from the river's confluence with the Columbia River. The area is famous for its natural beauty and recreational activities such as whitewater rafting and fishing. In 1996, the federal government ordered the operator (PacifiCorp) to alter the dam and add fish ladders to meet environmental codes. PacifiCorp deemed the modifications too expensive and asked for permission to instead decommission the dam. Decommissioning was approved and the dam was breached in October 2011 (according to Wikipedia). Taking out the dam is expected to reopen about 53 km of habitat for steelhead trout and about 22 km for Chinook salmon, depending on how well different runs of fish contend with natural falls in the river.

These projects also differ from the "-2,+1"-proposal. In the case of the Elwha and Glines Canyon Dams, the government and hence the taxpayers bear the cost of dam removals and restoration. So even if the shareholders in the hydropower company and those deriving use and/or nonuse values from the project gain (or at least don't lose), there might be those who lose from the project by having to pay without deriving enough values from the project. The Condit Dam was deemed too expensive to operate given new environmental standards. The choice of baseline or counterfactual is important in assessing this project. The operator loses if the baseline is the situation before environmental standards were altered. If the baseline is with the new standards the best the operator could do is to close down operations. In any

[3] The research team includes David Chapman, Rich Bishop, Jim Boyd, Colleen Donovan, John Duffield, Anthony Dvarskas, Megan Lawson, and John Loomis.

case, the cost of the new standards seems to be the value of electricity foregone as the dam is removed but there are also benefits, similar in a qualitative sense to those derived from removals of the two final dams in Ljusnan River.

In closing, we believe that our Win-Win approach could be used to analyze re-regulation in rivers in every part of the world, whether developed or not. An advantage is that it avoids monetary transfers or side payments that could be extremely difficult to design and handle. Instead it designs a package or set of physical measures. We are convinced that the same approach can be used to handle other sharp resource use conflicts. Just to take a simple example: If a new railway is constructed it might destroy wetlands that is a valuable bird rookery. Not only birds but also bird spotters are affected. However, it might be possible to construct new wetlands some miles away that both birds and bird watchers find at least as good as the initial area. Then a Win-Win situation has been created. This simple example illustrates that flexible approaches to the handling of sharp resource use conflicts might be fruitful.

References

1. Johansson, P.-O., Kriström, B.: The Economics of Evaluating Water Projects: Hydroelectricity Versus Other Uses. Springer, Heidelberg (2012)
2. BushBroker® Landowner–Frequently asked questions. Information sheet number 2 Department of Sustainability and Environment, Victoria, Austriala. http://www.dse.vic.gov.au/__data/assets/pdf_file/0011/149636/BB-info-2-FAQ-LO.pdf (2009)
3. Bayon, R.: Making Environmental Markets Work: Lessons from Early Experience with Sulfur, Carbon, Wetlands, and Other Related Markets. Forest Trends, Washington, D.C. (2004). Technical Report
4. Duffield, J.: The political economy of hydropower and fish in the Western US. In: Johansson, P.-O., Kriström, B. (eds.) Modern Cost-Benefit and Hydropower, pp. 100–120. Edward Elgar, Cheltenham, UK (2011)
5. Mansfield, C., Van Houtven, G., Hendershott, A., Chen, P., Porter, J., Nourani, V., Kilambi, V.: Klamath River Basin restoration: nonuse value survey (2012). Prepared for US Bureau of Reclamation

Appendix A
A Simple General Equilibrium Cost-Benefit Rule

In this Appendix we derive a very simple general equilibrium cost-benefit rule for a *small* project (and what we mean by a small project is clarified in what follows). The idea is to illustrate the flavor of such rules. The reader is referred to [1] for a detailed derivation of a rule for an economy with different taxes, spot markets for electricity, international trade, foreign ownership of key firms, energy certificates, markets for tradeable emission permits, and transmission of electricity modeled as a natural monopoly.

Let us assume an extremely simplified single-household economy (or an economy with a number of identical households or a society where the welfare distribution is more or less optimal) in order to set aside distributional considerations (but see Appendix B). The social welfare function, i.e. the indirect utility function, of this society is written as follows:

$$V(p, w, \pi - T, E) \tag{A.1}$$

where p is the price of a commodity, the price of the numéraire commodity is suppressed, w is the wage rate, π denotes profits of the price-taking firm producing the considered commodity, T is a lump-sum tax, and E is (unpriced and from the point of view of the household exogenous) environmental quality. The firm producing the commodity is assumed to use homogenous labor as the sole input. Its profits is stated as follows:

$$\pi = p \cdot x^s - w \cdot L^d \tag{A.2}$$

where x^s denotes supply of the good and L^d demand for labor.

The government is assumed to use labor to produce environmental quality. Its budget constraint is written as follows:

$$T = w \cdot L^g \tag{A.3}$$

P.-O. Johansson and B. Kriström, *Evaluating Water Projects*, SpringerBriefs in Economics, DOI: 10.1007/978-3-642-36790-8,

where L^g denotes its purchase of labor. We ignore any production function linking labor to the production of the environmental commodity.[1]

Let us assume that the economy is in a general equilibrium. Denote initial equilibrium levels by a superscript 0 (and use the same superscript to denote initial levels of the parameters (e.g. environmental quality) of the economy). Then a parameter, let us say E is changed from E^0 to E^1. This project moves the economy to a new general equilibrium. Denote the new equilibrium levels by a superscript 1. Assume that the social welfare function $V(.)$ is (at least) once continuously differentiable on its domain. Then the change in social welfare can be stated as follows:

$$\begin{aligned}\Delta V &= V(p^1, w^1, \pi^1 - T^1, E^1) - V(p^1, w^0, \pi^0 - T^0, E^0) \\ &= V_p \cdot \Delta p + V_w \cdot \Delta w + V_m \cdot (\Delta \pi - \Delta T) + V_E \cdot \Delta E + R\end{aligned} \tag{A.4}$$

where $V_p = \partial V(.)/\partial p$, $V_w = \partial V(.)/\partial w$, $V_m = \partial V(.)/\partial \pi = \partial V(.)/\partial T$, $V_E = \partial V(.)/\partial E$, R is called the remainder term, and all derivatives are evaluated at *initial* general equilibrium levels (i.e. at p^0, w^0, E^0 and so on).

The remainder term in Eq. (A.4) can be stated as $R(P^0, P^1)$, where P^0 (P^1) is a vector containing initial (final) values of prices and so on. A *small* project is here defined as a project such that we reasonably can argue that $R \approx 0$.[2] We might then as well look at the total differential dV instead of ΔV but for simplicity we will continue to use the Δ-symbol.

Set $R = 0$ and employ some of the first-order conditions for utility maximization. Then the change in utility, converted to monetary units by multiplication by the inverse of the marginal utility of lump-sum income (denoted λ) can be stated as follows:

$$\begin{aligned}\frac{\Delta V}{\lambda} &= -x^d \cdot \Delta p + L^s \cdot \Delta w + (\Delta \pi - \Delta T) + \frac{V_E}{\lambda} \cdot \Delta E \\ &= (x^s - x^d) \cdot \Delta p + (L^s - L^d - L^g) \cdot \Delta w \\ &\quad + p \cdot \Delta x^s - w \cdot \Delta L^d \\ &\quad + \frac{V_E}{\lambda} \cdot \Delta E - w \cdot \Delta L^g\end{aligned} \tag{A.5}$$

Let us briefly discuss some of the terms in the different lines of this equation. In comparison to the final line of Eq. (A.4) the first two terms in the first line of Eq. (A.5) are obtained by using first-order conditions for utility maximization: $V_p = -\lambda \cdot x^d$ and $V_L = \lambda \cdot L^s$ respectively (and multiplying by $1/\lambda$). The term within parentheses in the first line is obtained by noting that $V_m = \lambda$. In the second

---

[1] T could be considered as a residual since once E and hence L^g are fixed, the magnitude of the lump-sum tax follows.

[2] R has the following property: $\lim_{P^1 \to P^0} \left(\frac{R(.)}{\| P^0 - P^1 \|^2} \right) = 0$, where $\| P \|$ is the Euclidean norm of the vector P; see, for example, [2] for details.

line we have employed the fact that $\Delta\pi = \Delta p \cdot x^s + p \cdot \Delta x - \Delta w \cdot L^d - w \cdot \Delta L^d$ and a similar decomposition of ΔT using Eq. (A.3). In the two final lines we have collected the remaining terms.

The two terms in the second line of Eq. (A.5) are both equal to zero since supply is equal to demand for all commodities in a general equilibrium. Thus a *general equilibrium* assessment of our small project can be stated as follows:

$$\frac{\Delta V}{\lambda} = \frac{V_E}{\lambda} \cdot \Delta E - w \cdot \Delta L^g + p \cdot \Delta x^s - w \cdot \Delta L^d \tag{A.6}$$

The direct benefits of the project is the willingness to pay (WTP) for the improved environmental quality as covered by the first term. The second term covers the direct costs for the project. Both terms are evaluated at initial prices. Then there are two terms related to the change in profits of the firm producing commodity x. If the firm is a price-taking profit maximizer and the change is small, the two terms net out due to the first-order conditions for profit maximization: $p \cdot \Delta x^s = (p \cdot \partial f(L^d)/\partial L^d) \cdot \Delta L^d = w \cdot \Delta L^d$, where $f(.)$ is the production function.

However, in a hydropower context there might be a relationship between water use and environmental quality. For example, water might be diverted from hydropower generation to a dryway in order to allow fish migration. Therefore the profits change expression might reflect changes in production that are implemented in order to achieve the change[3] in E. Still, the general equilibrium cost-benefit rule for a small project is remarkably simple. In addition, the right-hand side of Eq. (A.6) is proportional to the *unobservable* change in utility ΔU in the left-hand side of the equation. Thus if a good estimate of the right-hand side expression is available and it has a positive (negative) sign we know that the considered project increases (decreases) utility/welfare.

The general equilibrium cost-benefit rules derived in [1] are much more general than the one stated in Eq. (A.6) since the context is more complex, involving different taxes, market power in transmission, and foreign ownership of the evaluated hydropower firm. Still, Eq. (A.6) gives a flavor of what a general equilibrium cost-benefit rule typically looks like. In fact, the major difference between a single-period version of the rule in [1] and the one in Eq. (A.6) is that the former contains a term reflecting the effects of distortionary taxes.

[3] Unless there are goodwill reasons for deviating from a profit maximizing strategy the firm must be compensated. For a detailed account the reader is referred to [1].

Appendix B
On a Win-Win Situation and the (Strong) Pareto Criterion

In this Appendix we identify a number of stakeholders in the second project, i.e. $-2, +1$. These are shareholders in the hydropower firm, those living in the catchment area, construction workers, taxpayers in other parts of the country than the Ljusnan catchment area, and a group consisting of individuals that might have altruistic reasons for supporting the project. For notational simplicity, let us assume that all prices remain unaffected by the considered project. For reasons of simplicity we also assume that the only tax is a lump-sum one.

Let us first consider a shareholder in the hydropower firm (Fortum), owning a share α^c. Such an individual is assumed to be no worse off with the project if it holds that:

$$V^c\left(p, w^c, \alpha^c \cdot \pi^1 - T^{c^1}\right) \geq V^c\left(p, w^c, \alpha^c \cdot \pi^0 - T^{c^0}\right) \tag{B.1}$$

where a superscript c refers to such a person ($c \in S^c$ and S^c is the set of such individuals living in Sweden), π^1 (π^0) refers to the hydropower firm's present value profits with (without) the project, and T^{c^1} (T^{c^0}) denotes tax payments with (without) the project. Basically this assumes that the firm's present value profits are at least as high with the project as without it, assuming that the individual's tax payments remain unaffected (or decreases). For simplicity, we assume that this individual is not concerned about environmental quality in the considered region of Sweden.

Consider next an individual living in Ljusnan's catchment area. Such an individual is assumed to be at least as well off with the project as without it if it holds that:

$$V^{lj}\left(p, w^{lj}, -T^{lj^1}, E^1\right) \geq V^{lj}\left(p, w^{lj}, -T^{lj^0}, E^0\right) \tag{B.2}$$

where a superscript lj refers to an individual living along Ljusnan River ($lj \in S^{lj}$). The main impact is through improved environmental quality, i.e. the change in E from E^0 to E^1. During construction of $+1$ and dam removals, the individual might be affected by disruptions such as noise and heavy transports but such disruptions

P.-O. Johansson and B. Kriström, *Evaluating Water Projects*, SpringerBriefs in Economics, DOI: 10.1007/978-3-642-36790-8,

are assumed to be covered by the E-parameter. One cannot completely rule out that the proposal would affect taxes. In any case, the individual is assumed to make an overall assessment of the proposal's consequences in the questionnaire.

Those involved in construction work and dam removals are assumed to live in the catchment area. Such a worker is assumed to be no worse off with the proposal if it holds that:

$$V^{cw}\left(p, w^{cw} \cdot L^{cw^1} - T^{cw^1}, E^1\right) \geq V^{cw}\left(p, w^{cw} \cdot L^{cw^0} - T^{cw^0}, E^0\right) \tag{B.3}$$

where a superscript cw refers to a construction worker ($cw \in S^{cw}$). For notational simplicity it is assumed that such a worker does not attribute value to leisure time explaining that labor income appears in the lump-sum income argument. However, the worker might still value the change in environmental quality as is reflected by the E-argument in his or her utility function. In any case, at least in theory, such an individual is in the sample and responds to the questionnaire.

A taxpayer living in another part of the country might be affected if the proposal changes taxes trough, say, changed profits taxes. This individual is a winner or at least not a loser from the project if it holds that:

$$V^t\left(p, w^t, -T^{t^1}\right) \geq V^t\left(p, w^t, -T^{t^0}\right) \tag{B.4}$$

where a superscript t refers to a taxpayer living in another part of the country ($t \in S^t$).

One cannot completely rule out that an individual living in another part of the country attributes value to the proposal because he or she cares about those living in the Ljusnan area. A pure altruist respects the preferences of others and gains from the proposal if it holds that:

$$V^{ta}\left(p, w^{ta}, -T^{ta^1}, V^{lja^1}\right) \geq V^{ta}\left(p, w^{ta}, -T^{ta^0}, V^{lja^0}\right) \tag{B.5}$$

where a superscript ta refers to the altruist ($ta \in S^{ta}$). This individual cares about the utility attained by those living in the catchment area, i.e. V^{lja}, which is interpreted as a vector with elements V^{lj}. Alternatively, an individual might be a paternalistic altruist in the sense that he or she cares about the wealth or the health or some other aspect of individuals' welfare. In addition, an individual might care about species living in the area. For details on different altruism concepts the reader is referred to [1].

If someone in the combined set (i.e. $S^c + S^{lj} + S^{cw} + S^t + S^{ta}$) of individuals is better off and no one is worse off with the proposal, the proposal satisfies the *strong* Pareto criterion[4]; see, for example, [3] or [4].

We conclude by noting that our approach here seems to resemble the one proposed by the Nobel laureate James Buchanan in an article from the late 1950s;

---

4 The *weak* version of the criterion requires that everyone is better off. Strong Pareto implies weak Pareto (but the converse is not necessarily true). However, under certain technical assumptions they are equivalent; see, for example, [4] or [6].

see [5]. Buchanan's proposal is that a project is desirable if it "results in (1) everyone being better off or (2) someone being better off and no one being worse off than before the change" (p. 125). This assumes that compensation is *actually* paid to those losing from a project. This is different from the typical cost-benefit analysis, which rather looks at *hypothetical* compensation in the sense that a project is desirable if those who gain are hypothetically able to compensate those who lose.[5] Thus one could apply this latter rule to the small project analyzed in Appendix A. However, there are also other approaches to cost-benefit analysis, for example those that explicitly weigh different individuals using a social welfare function (or some other weighting tool) as is further discussed in Chap. 4.

[5] Boadway did show that this approach holds only for small projects, in general; see [3] for a proof.

Appendix C
The Questionnaires

The questionnaire relating to the Dönje power plant is contained in [1] and is not replicated here. However, the central valuation questions are found in Sect. 2.3. The remaining questions are more or less identical to the socioeconomic ones in the questionnaire below.

The web-questionnaire relating to the "$-2, +1$"-proposal was commenced by a film that the respondent could skip. It gave the same information as the written information provided later on. Pictures showing what the river looks like today and would look like after removal of the two dams at the end of the river (in the short run and in the longer run when the environment had adjusted to the removals) were shown to the respondent. Then the following (here somewhat simplified) questions were asked.

1. Activities in Ljusnan during last year (Fishing, bathing, skating, walking, other)
2. Do you live in property close to Ljusnan?
3. Do you own property close to Ljusnan?
4. Re-regulation of Ljusnan as described as in Chap. 3. followed by a detailed presentation
5. What do you think of this proposal (in comparison to today's situation)?
    a. Definitely an improvement
    b. Perhaps an improvement
    c. Neither better nor worse
    d. Perhaps a worsening
    e. Definitely a worsening
    f. I don't know
6. Gender
7. Year of birth
8. Postal code
9. Highest education
10. Employment

P.-O. Johansson and B. Kriström, *Evaluating Water Projects*,
SpringerBriefs in Economics, DOI: 10.1007/978-3-642-36790-8,

11. Household income net of tax
12. # of children <18 years old in household
13. # of persons ≥ 18 years in household
14. # of persons ≥ 18 years that contribute to household income

References

1. Johansson, P.O., Kriström, B.: The Economics of Evaluating Water Projects. Hydroelectricity Versus Other Uses. Springer, Heidelberg (2012)
2. Sundaram, R.: A First Course in Optimization Theory. Cambridge University Press, Cambridge (1996)
3. Boadway, R.W., Bruce, N.: Welfare Economics. Basil Blackwell, Oxford (1984)
4. Mas-Colell, A., Whinston, M.D., Green, J.R.: Microeconomic Theory. Oxford University Press, Oxford (1995)
5. Buchanan, J.M.: Positive economics, welfare economics, and political economy. J. Law Econ. **2**, 124–138 (1959)
6. Jehle, G.A., Reny, P.J.: Advanced Microeconomic Theory. Addison-Wesley, Boston (2001)

Index

A
Altruism, 15, 23, 36, 45
Altruism, paternalistic, 23
Altruism, pure, 26

B
Baseline, 12, 37, 53
Belyaev-Kriström estimator, 20
Bollnäs, 4, 5, 13, 17–19, 25, 41

C
Carbon dioxide, 1, 16, 20
Carbon equivalent, 20
Certainty equivalent price, 22
Climate gases, 1, 14, 16, 20, 23, 26, 27
Compensation, 12, 15, 23, 26, 29, 49, 50
Compensation, actual, 61
Compensation, hypothetical, 50
Confidence interval, 27
Consumer sovereignty, 15
Contingent valuation, 16, 17, 25, 26
Cost-benefit acceptability curve, 30
Cost-benefit analysis (CBA), 1, 2, 8, 9, 11, 15, 16, 19, 20, 24, 27, 29, 41, 49, 50
Counterfactual, 12, 37, 46, 53

D
Dönje, 4, 12–14, 18, 21, 22, 27, 29
Discount rate, financial, 40
Discount rate, social, 23–25
Discounting, hyperbolic, 25
Dryway, 5

E
EcoSenseLE, 23, 27
EEX market, 21, 22
Egoism, nationalistic, 15
Elasticity of marginal utility, 24
Electricity tax, 29
Elwha River, 53
Emissions, 1, 14, 15, 23, 26, 27
Energy certificates, 16, 29
Euclidean norm, 56
European Union Allowance (EUA), 20
Externality, 16, 23, 26, 27, 30, 41, 49

F
Fishing benefits, 5, 17, 18
Fortum, 2, 7, 13, 18, 29
Future generations, 25, 26, 36, 45

G
General equilibrium, 1, 11, 12, 15, 16, 24, 27
General equilibrium model, computable, 23, 27

H
Hamiltonian, 24
Hicks–Kaldor, 3, 43, 50
Hydroelectric settlement agreement, 52
Hydropower, 1, 13, 22
Hydropower plant, 15, 16

P.-O. Johansson and B. Kriström, *Evaluating Water Projects*, SpringerBriefs in Economics, DOI: 10.1007/978-3-642-36790-8,

I
Income, 24
Indifference curves, 28
Investment, 24, 25, 38, 40
Isoquants, 28

J
Jammalamadaka estimator, 20

K
Klamath river, 52
Krutilla–Fisher argument, 46

L
Life expectancy, 26
Lignite-fired plant, 21
Ljusnan, 4, 35–38, 41, 42, 45, 50
Lump-sum tax, 55

M
Marginal Cost of Public Funds (MCPF), 16
Marginal Excess Burden of Taxes (MEB), 16
Marginal social utility of income, 50
Marginal utility of income, 50
Merit order curve, 21
Minimum flow, 5
Monte Carlo simulation, 29
Morbidity, 23
Mortality, 23

N
Natural monopoly, 16
Non-use value, 44, 45
Nord Pool market, 14, 20–22
Nuclear power, 13

O
Offsets, 51

P
Pareto criterion, strong, 60
Pareto criterion, weak, 60
Pareto principle, 7, 37, 43–45, 49, 50
Particulates, 23, 27
Payment format, interval, 19
Permits, 14–16, 20, 27
Pilot study, 18, 25
Present value, 12, 22, 24–27
Price elasticity, 28, 29
Price forecast, 21, 22, 27, 30
Project, small, 56
Pumped-storage plant, 21, 27

R
Ramsey model, 24
Rawlsian, 50
Recreational benefits, 1, 5, 13
Regulating power, 16
Remainder term, 56
Risk aversion, 24
Rivers, regulated, 1

S
Sensitivity analysis, 25, 27–29
Social profitability, 27, 30, 31
Social welfare function, 50
Spot price, 20–22, 29
Status quo, 37
Sulphur, 15, 26
System price, 20, 22

T
Time horizon, 26, 30
Time preference, 24
Tornado diagram, 28
Transmission, 16, 20–22, 29
Triangular distribution, 30
Turbines, 4, 21

U
U.S. Army Corps of Engineers (ACE), 2
Uniform distribution, 30
Unit tax, 29
Utility function, 12

V
Value added tax (VAT), 29
Vattenfall, 29
Voxnan, 3

W
Water Framework Directive (WFD), 1
Water rights, 18, 26, 53
Web survey, 11, 18, 40, 41, 45
Weibull distribution, 20

Welfare weight, 50
Wetlands, 51, 54
White Salmon river, 53
Willingness to accept compensation (WTA), 13
Willingness to pay (WTP), 13, 15, 19, 25, 26, 30
Win-Win, 3, 7, 9, 35, 45, 49, 50, 54

MIX
Papier aus verantwortungsvollen Quellen
Paper from responsible sources
FSC® C105338

If you have any concerns about our products,
you can contact us on
ProductSafety@springernature.com

In case Publisher is established outside the EU,
the EU authorized representative is:
Springer Nature Customer Service Center GmbH
Europaplatz 3, 69115 Heidelberg, Germany

Printed by Libri Plureos GmbH
in Hamburg, Germany